Practical Handbook of Thermal Fluid Science

Authored by

Yun Wang

Practical Handbook of Thermal Fluid Science

Author: Yun Wang

ISBN (Online): 978-1-68108-919-5

ISBN (Print): 978-1-68108-920-1

ISBN (Paperback): 978-1-68108-921-8

© 2023, Bentham Books imprint.

Published by Bentham Science Publishers – Sharjah, UAE. All Rights Reserved.

First published in 2023.

General:

1. Any dispute or claim arising out of or in connection with this License Agreement or the Work (including non-contractual disputes or claims) will be governed by and construed in accordance with the laws of the U.A.E. as applied in the Emirate of Dubai. Each party agrees that the courts of the Emirate of Dubai shall have exclusive jurisdiction to settle any dispute or claim arising out of or in connection with this License Agreement or the Work (including non-contractual disputes or claims).
2. Your rights under this License Agreement will automatically terminate without notice and without the need for a court order if at any point you breach any terms of this License Agreement. In no event will any delay or failure by Bentham Science Publishers in enforcing your compliance with this License Agreement constitute a waiver of any of its rights.
3. You acknowledge that you have read this License Agreement, and agree to be bound by its terms and conditions. To the extent that any other terms and conditions presented on any website of Bentham Science Publishers conflict with, or are inconsistent with, the terms and conditions set out in this License Agreement, you acknowledge that the terms and conditions set out in this License Agreement shall prevail.

Bentham Science Publishers Ltd.
Executive Suite Y - 2
PO Box 7917, Saif Zone
Sharjah, U.A.E.
Email: subscriptions@benthamscience.net

**BENTHAM
SCIENCE**

CONTENTS

PREFACE

Thermal fluid science is an important traditional subject in engineering, which deals with thermodynamics, fluid flow, and heat transfer. Tremendous efforts have been made so far, particularly during the last couple of centuries or so, on advancing the technology and fundamental research of heat engines and fluid flow and heat transfer devices. In addition to the large number of research and review paper publications, many classic books have been published and are available in the market, primarily for standard fundamental and classroom learning with few experimental and real-world data for learning and exercise. This book contributes to this aspect of thermal fluid science and technology; it focuses on experimental and real-world operating data and relevant fundamentals. For examples, operating data in power plants regarding boilers, steam turbines, and gas turbines are provided to analyze efficiencies, along with fundamentals and use of the steam tables. Experimental data of Venturi and orifice plate flow meters are provided to show step by step how to calibrate these two important flow meters, along with experimental steps. Detailed experimental data of wind tunnels, sphere heating/cooling, pipe flow, engines, and refrigerators/heat pumps are provided to show how to test and use these important devices and how to evaluate the heat transfer coefficients and friction coefficient/factor. Figures from original patents are directly used to show readers how to prepare figures in patents. Useful data, equations, solutions, and correlations are given in the Appendix sections.

More specifically, this book volume aims to relate the thermal-fluid science fundamentals with real-world operation of important devices that greatly impact our daily lives, such as power generation, heat transfer, air conditioning, refrigeration, engines, flow meters, airplane flying, and pipe flows. In other words, the objective of this book is to provide an introduction to the essential knowledge required to perform analysis and evaluation for practical systems and several major inventions; the book also presents and discusses the experimental methods and apparatus. In addition to providing an introduction in Chapter 1, major concerns in the thermal fluid laboratory, such as safety and training, are summarized. The book outlines the basic methods for data statistics and error analysis in Chapter 2, along with an experiment method and data for exercise. Chapter 3 of the book presents the fundamentals

of heat transfer, measurement of temperature and heat transfer coefficient, and an example of sphere cooling/heat experiment. Following the format of Chapter 3, Chapters 4-9 describe and discuss power plant operation, pipe flow, and flow meters, power plant efficiencies, wind tunnel, engines, and refrigeration. Chapter 10 focuses on the report preparation and basics of dimensions, units, and significant figures, along with the requirement for figures and graphs.

This book is based on my about 15 years of teaching experience at the UC Irvine MAE107 Thermal Fluid Science Laboratory. I would like to thank Daniel Kahl, Bongjin Seo, Qin Chen, Jingtian Wu, Patrick Hong, Frederick R. Bockmiller, and Daniela Fernanda Ruiz Diaz at the UC Irvine for their assistance in the book preparation.

CONSENT FOR PUBLICATION

Not applicable.

CONFLICT OF INTEREST

The author declares no conflict of interest, financial or otherwise.

ACKNOWLEDGEMENTS

Declared none.

Yun Wang

SYMBOLS

A:	Area [m^2]; Surface
A_s:	Surface area [m^2]
c:	Standard deviation of the wind speed [m/s]; Specific heat per mass [J/K kg]
C:	Specific heat [J/K] or [J/K kg]
C_D:	Drag coefficient [-]
C_o:	Orifice coefficient [-]
C_v:	Specific heat at constant volume [J/K] or [J/K kg]; Venturi coefficient [-]
C_p:	Specific heat at constant pressure [J/K] or [J/K kg]
D:	Diameter [m]; Cross section diameter [m]
e:	Internal energy [J]; potential [V]
E:	Radiation heat flux [J/m^2/s]; Total Energy [J] or [BTU]
f:	Fanning friction factor [-]
F:	Force [N]
h:	Height [m]; Heat transfer coefficient [W/(m^2·K)]; Specific enthalpy [J/kg]
h_{vl}:	Latent heat of water condensation [J/kg] or [BTU/scf]
ΔH:	Liquid height [m]
k:	Mean wind speed [m/s]; Ratio of specific heats [-]; Thermal conductivity [W/(m·K)]
K:	Calibration constant [K/V] or [^{o}C/V]
L_w:	Length [m]
m:	Mass [kg]
p:	Pressure [N/m^2]
P:	Pressure [N/m^2]; Probability [-]
q:	Heat flux [J/m^2/s]
$\dot{q}$:	Energy [J] or [BTU]
q'':	Heat flux [J/m^2/s]
Q:	Surface heat flow rate [W]
ΔQ:	Heat added to the system [J]
$\dot{Q}$:	Volume flow rate [m^3/s]
r:	Compression ratio [-]; Radial dimension
R:	Radius [m]
Re:	Reynolds number [-]

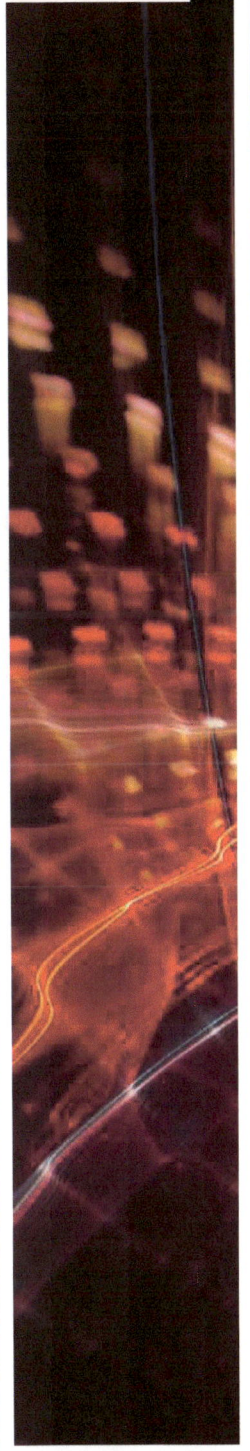

R_T:	Resistivity [K/W]
s:	Entropy [J/K]
t:	Time [s]
T:	Temperature [K] or [°C]
T_∞:	Ambient temperature [K]
T_s:	Surface temperature [K]
T_{surr}:	Surrounding temperature [K] or [°C]
u:	Velocity [m/s]
$\bar{u}$:	Average velocity [m/s]
U:	Velocity [m/s]
$\bar{U}$:	Mean velocity [m/s]
v:	Specific Volume [m³/kg]
V:	Volume [m³]
$\dot{V}$:	Volume flow rate [m³/s]
w:	Work [J] or [J/kg]
W:	Work [J]
ΔW:	Work done to the surroundings [J]
$\dot{W}$:	Work flow rate [W]

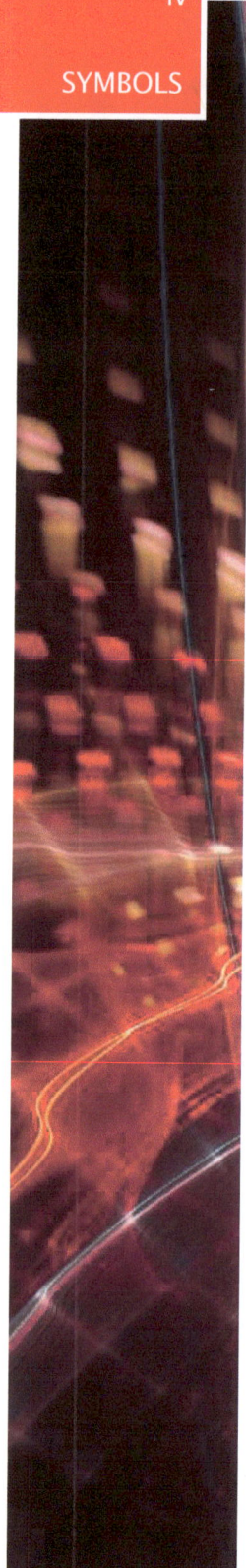

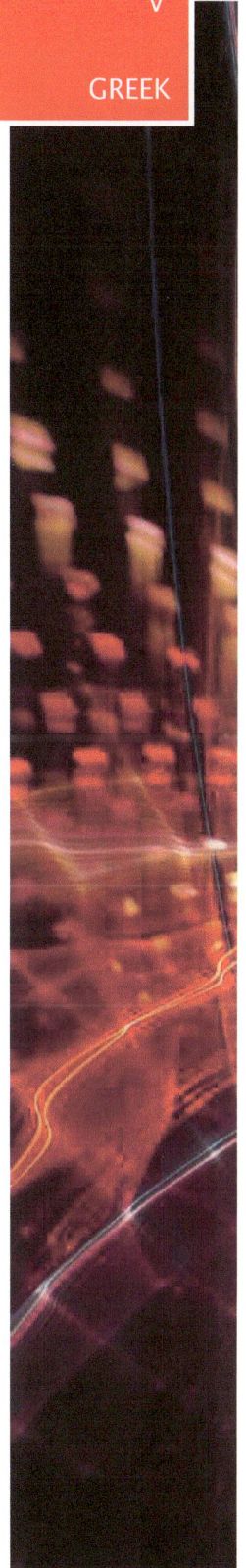

GREEK

α: Thermal diffusivity [m^2/s]

β: Ratio of diameter [-]

ϵ: Roughness factor [-]; Emissivity [-]

Θ: Dimensionless temperature [-]

η: Thermal efficiency [-]

μ: Dynamic viscosity [Pa·s]

ρ: Density [kg/m^3]

σ: Stress tensor [N/m^2]; Boltzmann constant [5.6703x10^{-8} W/m^2/K^4]

τ: Time constant [-]

τ_w: Viscous shear stress [N/m^2]

υ: Kinematic viscosity [m^2/s]

INTRODUCTION

1.1. INTRODUCTION TO THERMODYNAMICS, FLUID FLOW, AND HEAT TRANSFER

Thermodynamics, fluid flow, and heat transfer not only play an important role in science and engineering but also in everyday life. For example, gasoline vehicles operate on the Otto cycle, residential air-conditioners are developed based on the vapor-compression refrigeration cycle, steam turbines are the central unit in traditional power plants, and flow meters and pipe flows are inherent in both industrial and residential applications.

Thermodynamics can be defined as the science of energy, which deals with heat and temperature and their relationship with energy, work, and properties of matter. The knowledge about thermodynamics behaviors comes from observations, which are then formulated into laws, including the well-known first and second laws of thermodynamics [1]. Heat, work, entropy, various types of energy, efficiency, pressure, force, and temperature are the major metrics used when studying thermodynamics.

Heat transfer relates to the generation, use, conversion, and exchange of heat or thermal energy. Heat is transferred by three main modes, including heat conduction, convection, and radiation. In addition, phase change involves latent heat release or absorption. In general, thermodynamics describes equilibrium states, *i.e.* state properties such as temperature, pressure, and internal energy are the same in all the spatial dimensions of a system. Heat transfer, in contrast, depicts non-equilibrium phenomena such as temperature gradients. The popular heat transfer properties include thermal conductivity, heat transfer coefficient, thermal diffusivity, and emissivity. Temperature gradients and evolution and heat flux/rates are the major metrics used when studying heat transfer.

Fluid mechanics describes the flow of fluids, such as liquid, gas, and plasma. It describes basic concepts and governing equations underlying fluid behaviors, including the conservation principles of mass and momentum. Viscosity is an important concept relating to viscous flows. Shockwaves occur when the Mach number reaches 1 or the fluid velocity approaches the sound speed. Turbulent flows are frequently encountered in nature and industrial applications. The distributions and evolutions of fluid velocity, density, and

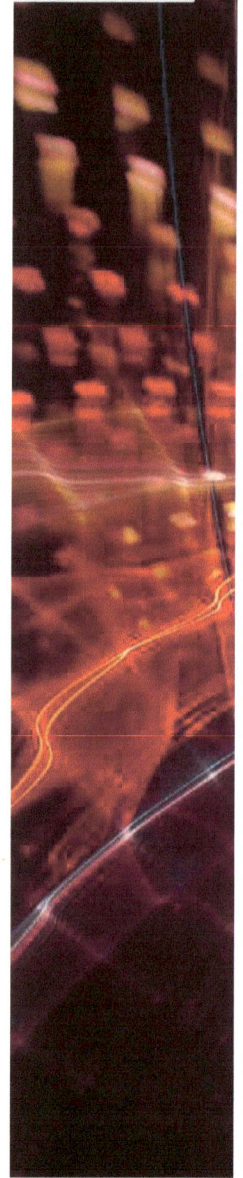

pressure, and flow rate are the major metrics used when studying fluid mechanics. Table **1.1** summarizes the major equations and laws in the three subjects.

Table 1.1 Typical equations in thermodynamics, heat transfer, and fluid mechanics [2, 3].

Thermodynamics	1st Law: $\Delta E = Q - W$ 2nd Law: $ds \geq \frac{\delta Q}{T}$
Heat Transfer	$\rho \left[\frac{\partial e}{\partial t} + \boldsymbol{u} \cdot \nabla e \right] = \nabla \cdot (\mathrm{k}\nabla T) + \mu \boldsymbol{\Phi} + \dot{q}$
Fluid Mechanics	Continuity equation: $\frac{\partial \rho}{\partial t} + \nabla \cdot (\rho \boldsymbol{u}) = 0$ Momentum Equation: $\rho \left[\frac{\partial \boldsymbol{u}}{\partial t} + \boldsymbol{u} \cdot \nabla \boldsymbol{u} \right] = -\nabla \mathrm{p} +$ $\nabla \cdot [\mu(\nabla \boldsymbol{u} + (\nabla \boldsymbol{u})^T)] + \rho \boldsymbol{g}$

1.2. EXPERIMENTAL MEASUREMENT

Measurement is frequently encountered in daily activities. For example, rulers and body thermometers are used to measure a person's height and body temperature. Gas stations need to quantify how many gallons or liters of gasoline are added to a fuel tank. Experimental measurement is not merely reading a number and obtaining a value. The selection of measurement equipment/apparatus, techniques, and methods to meet any accuracy requirements is an important task for any experimental work. In practice, direct measurement is not always the final target. For example, in a liquid-in-glass thermometer, the temperature is not directly measured. Instead, the liquid length is measured by a ruler, which is then converted to temperature through their correlation. In a thermocouple, the voltage difference between the two junctions is directly measured by a voltage meter, which is then converted to temperature through their relationship. Any errors in the direct measurement of length or voltage using a ruler or voltage meter will eventually pass to the final temperature measurement. Thus, understanding error propagation from direct measurement to the final value, *i.e.*, error or uncertainty analysis, is important for experimental design, equipment selection, and method development.

In thermal-fluid experiments, metrics such as time, mass, volume, length, velocity, flow rate, voltage, temperature, pressure, and current are frequently used. Various types of equipment have been developed to measure these quantities. Each of them has a specific range of applications with corresponding precision and full range. In general, a high-precision measuring instrument is costly, and budget plays an important role in equipment selection and project design, especially in industrial development. For example, a school ruler of 1 mm in precision, which suffices for K-12 work, costs about one dollar. A micrometer caliper of 0.001 mm precision, widely used in engineering and scientific work, costs about 10-100 dollars. Table **1.2** lists a few examples of equipment for temperature, pressure, velocity, or flow rate measurement.

Table 1.2. Equipment for the thermal-fluid experiment.

Quantity	Instrument	Information	Image
Temperature	Thermocouples	Measure the temperature-dependent voltage based on the thermoelectric effect.	
	Liquid-in-glass Thermometers	Use liquid thermal expansion.	
	Infrared sensors	Measure the radiation.	
	Bimetallic devices	Use a bimetallic strip to convert temperature to mechanical displacement.	
Pressure	Liquid column elements	Use hydrostatics.	
	Elastic element gauge	Use elastic measuring elements.	
	Electrical transducers	Measure the deformation of elastic material under pressure and convert it to pressure.	

(Table 1.2) cont.....

Velocity	Pitot tube	Use the Bernoulli equation.	
	Hot-wire anemometers	Measure heat loss of the wire placed in a fluid stream.	
	Laser Doppler velocimeter	Use the Doppler shift in a laser beam to measure the velocity in transparent or semi-transparent fluid flows.	
Flow Rate	Differential pressure flow meters	Measure pressure drop and convert to flow rate using the Bernoulli equation.	
	Positive displacement flow meters	Measure the fluid volume in a given time period.	
	Velocity flow meters	Measure the stream velocity to calculate the volumetric flow rate.	
	Thermal mass flow meters	Measure the total mass flow rate.	

1.3. EXPERIMENTAL TEST PLAN AND STANDARD OF PROCEDURE

An experimental test plan is important for any experimental task. Figliola and Beasley [4] introduced three steps for an effective test plan, as below:

1.) "Parameter design plan: Determine the test objective and identify variables and parameters and a means for their control. Ask: What questions am I trying to answer? What needs to be measured? What factors may affect my results."

2.) "System and tolerance design plan: Select a measurement technique, equipment, and test procedure based on some preconceived tolerance limits for error. Ask: In what ways can I do the measurement, and how well do the results answer my questions?"

3.) "Data reduction design plan: Plan how to analyze, present, and use the anticipated data. Ask: How will I interpret the resulting data? How will I use the data to answer my question? How good is my answer? Does my answer make sense?"

In the next chapters, a series of testing samples will be introduced, following the three steps to describe and explain each experiment.

Additionally, standard operating procedure (SOP) is widely used in the professional world. An SOP consists of step-by-step instructions to help experimentalists conduct complex routine operations or experiments. The main purposes of an SOP are to improve operation efficiency, experimental quality, and uniformity of equipment use and to avoid miscommunication and failure to comply with regulations. A practical example of an SOP is a food recipe, which lists steps in making a specific dish, including ingredients, fire level, cook time, smell, and taste. Instruction for assemblies, such as IKIA's furniture and LEGO products, also serves as a simple SOP example.

In engineering and scientific laboratory or industry operation, SOPs need strict regulatory requirements. Example contents of such SOPs may include:

Purpose and Scope	Monitoring and Safety Systems
Responsibility	Waste Disposal/Cleanup
Materials and Equipment	Emergency Response Plan Procedure
Definitions	References
Specific Safety and Environmental Hazards	Preventive maintenance
Hazard Control	Monitoring and Safety Systems
Location of nearest emergency safety equipment	Emergency Response Plan
	References
Shipping and Receiving Requirements	Training Requirement
Step-by-step Operating Procedure	Additional Notes and Attachments
Special handling procedures, transport, and storage requirements	Documentation of Training
Preventive Maintenance	

Appendix VII shows a template for UC Irvine's SOP used in scientific experiment and operation.

1.4. EXPERIMENT AND LABORATORY SAFETY

1.4.1. Safety

An experiment is conducted to produce training, products, learning, or knowledge. Given that hands-on work and the use of various chemicals and tools are usually involved, there are risks and safety concerns in conducting experiments. In general, a laboratory is not designed for activities such as sleeping, playing, cooking, holding a party, or having food. Experiments can be dangerous if the safety guidelines and rules are not followed. Because toxic materials, fire heating, or the high voltage may be used or stored in a laboratory, facilities need to be equipped properly with safety equipment to provide timely protection from unexpected accidents, such as first-aid, fire distinguisher, and shower. Fatal accidents can occur in laboratories. For example, a research assistant was seriously burned in a UCLA laboratory fire on December 29, 2008, and consequently died of her injuries the next month. The EH&S (Environmental Health and Safety) investigation reports that the accident occurred while the assistant was working with T-Butyl lithium, a highly flammable compound. "She was allegedly wearing a synthetic material sweater and not a lab coat. Synthetic clothing can often burn readily and vigorously. The extent of injuries could probably have been reduced by quickly getting under the nearby safety shower rather than having a lab mate extinguish the flames." [11] Table **1.3** lists different types of laboratory accidents at different universities in the past.

#Example 1: Describe dress code for a stream-turbine power plant field trip.

Solution (suggestions):

There are many high-temperature pipes or vessels, such as steam pipes, boilers, and manifolds, to produce or deliver steam in a power plant. In most cases, long pants, closed toe & closed heel shoes, and sleeved shirts are required in a field trip. Helmets or safety goggles are recommended and required in some sites where falling or moving objects may cause injures.

Table 1.3. Several laboratory accidents in universities.

Location	Description of Accident	Year	Reference
UCSB	HF Lecture Bottle Explosion	-	[5]
UCSB	Rupture of Vacuum Flask	-	[6]
UT Austin	Chemistry Lab Fire	1996	[7]
UCSC	Fire Caused by Chemical	2000	[8]
UCI	Benzene Vapor Explosion	2001	[9]
OSU	Fighting Lab Fires	2005	[10]
UCLA	Lab Fire Fatality	2008	[11]
Yale	Chemistry Lab Accident	2011	[12]
University of Missouri	Anaerobic Chamber Explosion	2012	[13]
University of Hawaii	Hydrogen/Oxygen Explosion	2016	[14]
University of Maryland	Chemistry Lab Fire	2019	[15]

1.4.2. Personal Protective Equipment

Personal protective equipment (PPE) is important to lab work safety, which includes clothing and work accessories designed to protect people from workplace hazards. Lab coats, protective gloves, and safety goggles/shields are the basic PPE equipment used in a laboratory. A lab coat should be worn whenever thermal, chemical, or biological materials are handled. Lab coats protect the wearer's personal clothing and exposed body, such as arms and neck, from contaminants or hot liquids. Gloves should be worn whenever handling hazardous or high/low-temperature materials. There are various types of PPE gloves, such as chemical resistant gloves, heat resistant gloves, *etc*. Note that chemical-resistant gloves may protect against a few chemical hazards. Thus, carefully reviewing the material data sheet is important to glove selection. Eye and face protection is extremely important when hazards could

cause injury. Safety goggles protect against mechanical impact or projectile injures. Chemical splash goggles protect against liquid spills. Special types of goggles need to be worn when handling laser or UV hazards. Face shields are often used to protect the entire face. Safety glasses or goggles should be worn under face shields. For educational laboratory work, clothing choices play an important role in personal protection. Personal clothing offers a measure of protection against chemical or hot water splash and other hazards. Long pants are usually required in a laboratory or hands-on work environment, which provides much better protection than shorts or short skirts. Closed-toe shoes are required for lab work in many environments, including field trips, which protect visitors from injury due to chemical spills, moving machinery, sharp materials, hot objects, and falling solids. Flip Flops, Crocs, or other shoes with holes in the tops should be avoided, which provides limited foot protection. It is also recommended to secure loose clothing and tie back long hair to avoid being caught and dragged through chemicals in beakers, open flames, or rotating shafts. Fig. (**1.1**) shows the PPE signs for laboratory work.

Fig. (1.1). PPE signs [16].

#Example 2: Recommend PPEs for experiment using: a.) soldering; b.) liquid nitrogen; and c.) UV lights.

Solution (suggestions):
a.) Soldering requires the use of protective eye wear.
b.) Using liquid nitrogen requires the use of a lab coat, cryogenic gloves, protective eye wear, and a face shield.
c.) Ultraviolet (UV) radiation can adversely affect health depending on the duration of exposure and the UV wavelength, including erythema (sunburn), photokeratitis (a feeling of sand in the eyes), skin cancer, melanoma, cataracts, and retinal burns. The National Institute for Occupational Safety and Health (NIOSH) recommends that the exposure duration to 100 microwatts/cm² UV at 254 nm wavelength not exceed 1 minute. Workers exposed to harmful amounts and wavelengths of UV must take steps to shield themselves and limit exposure duration. Appropriate PPE includes gloves, a lab coat with no gap between the cuff and the glove, and a UV resistant face shield for eye and skin protection. Note there are three UV ranges: UV-A (314-400 nm), UV-B (280-315 nm), and UV-C (100-280 nm), which affect health differently.

1.4.3. Laboratory Safety Training

Before performing any lab tasks, proper laboratory safety training is required. Fig. (**1.2**) shows the major components in lab safety. In standard lab facilities, all participants must undergo laboratory safety training relevant to the facilities and experimental tasks. Training includes gaining knowledge and awareness to identify hazardous materials in the laboratory space, learn handling procedures, and take protective measures while working. For example, laboratories equipped with high-pressure cylinders require lab researchers to take High-pressure Gas Safety training, which introduces the basics of cylinders and handling procedures. Lab work that involves high-voltage power or equipment requires Arc Flash Training. Fig. (**1.3**) shows examples of hazardous communication standard pictograms. In addition, basic training, annual refresher training may be required for employees and supervisors as well as faculty, staff, and students. Note that laboratories that conduct R&D activities and related analytical work are subject to the requirements of the laboratory safety standard, even though they are used only to support manufacturing. Table **1.4** lists a few examples of Laboratory Safety training.

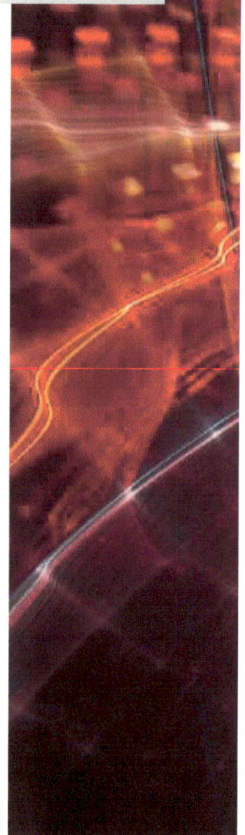

Table 1.4. Examples of laboratory safety training.

Lab Safety Training	Purpose & Description	References
Fire safety training	Fire prevention; Emergency preparedness and evacuation.	[17]
High-pressure gas training	Description of compressed gas cylinders; Operate on compressed gas cylinders; Potential hazards of compressed gas cylinders; Practice for storage, transport, and use.	[18]
High voltage training	Regulations & Training; High voltage hazards; Work practices & Responsibilities; HV safety equipment & Use; Rules & Policies for HV safety; Mobile equipment – HV hazards; HV exercise, *etc.*	[19]
Biological safety training	Comprehensive integrated training for all high-containment laboratory personnel groups	[20]
Chemical safety training	Basic chemical and lab safety awareness training	[21]
Radiation safety training	Basic training for new radioactive materials or X-ray machines and/or particle accelerator users.	[22]

#Example 3: Recommend lab safety training for an experiment using a.) flammable substance; b.) nuclear substance; and c.) drilling/milling machines.

Solution (suggestions):

Based on UCLC (UC Learning Center), the below trainings are recommended:
a.) Fire Safety and Hazardous Waste Training. Also see the UCLA accident in 2008. Chemical Safety Training is helpful.
b.) Radiation Safety, Hazardous Waste, and Hazardous Materials Incidents Emergency Procedures Training.
c.) Shop Safety Training.
In addition, Lab Safety Fundamentals Training is required for any laboratory work. Note that other university or laboratory systems may have different names for the training.

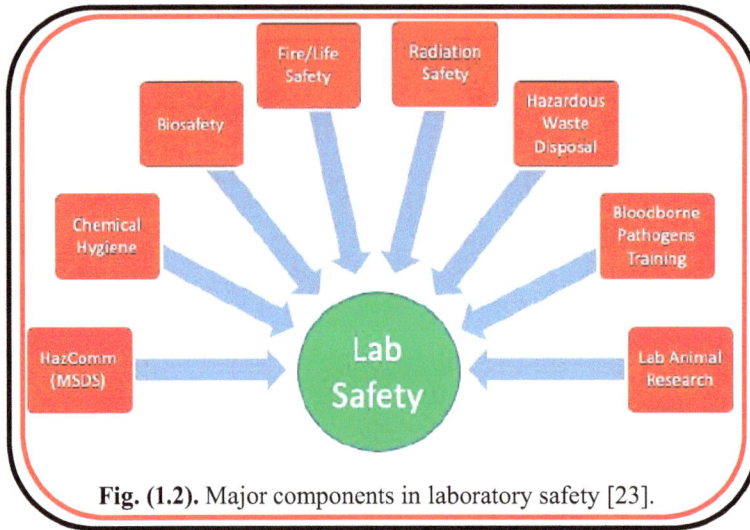

Fig. (1.2). Major components in laboratory safety [23].

Health Hazard	Flame	Exclamation Mark
• Carcinogen • Mutagenicity • Reproductive Toxicity • Respiratory Sensitizer • Target Organ Toxicity • Aspiration Toxicity	• Flammables • Pyrophorics • Self-Heating • Emits Flammable Gas • Self-Reactives • Organic Peroxides	• Irritant (skin and eye) • Skin Sensitizer • Acute Toxicity (harmful) • Narcotic Effects • Respiratory Tract Irritant • Hazardous to Ozone Layer (Non-Mandatory)
Gas Cylinder	**Corrosion**	**Exploding Bomb**
• Gases Under Pressure	• Skin Corrosion/ Burns • Eye Damage • Corrosive to Metals	• Explosives • Self-Reactives • Organic Peroxides
Flame Over Circle	**Environment** (Non-Mandatory)	**Skull and Crossbones**
• Oxidizers	• Aquatic Toxicity	• Acute Toxicity (fatal or toxic)

Fig. (1.3). Hazard communication standard pictogram [24].

1.5. QUESTIONS

1. What are PPEs?

2. List major steps in boiling water.

3. List major steps in frying eggs.

4. Write 1st law thermodynamics.

5. Write 2nd law thermodynamics.

6. Please list three types of goggles with their protection in lab work.

7. Please list three types of PPEs with their protection in lab work.

8. List three types of safety training and explain them briefly.

9. Please sketch the "gases under pressure" pictogram.

10. List 3-4 components in lab safety.

11. What is SOP?

12. Identify anything that looks wrong in a lab in Fig. (**S1.1**) [25].

13. One needs to use hydrogen cylinders in testing, list relevant lab safety training.

14. What is the major difference between thermodynamics and heat transfer?

15. For PEM fuel cell experiment, list the relevant types of lab safety training.

16. In battery fabrication facilities, what in hazard communication standard pictogram should be posted?

17. State three laws of Thermodynamics. For each, give an application in everyday life.

18. What are the three modes of heat transfer? For each, give an application in everyday life.

19. What is the Reynolds number and what is its significance? Give an application in everyday life for different Reynolds numbers.

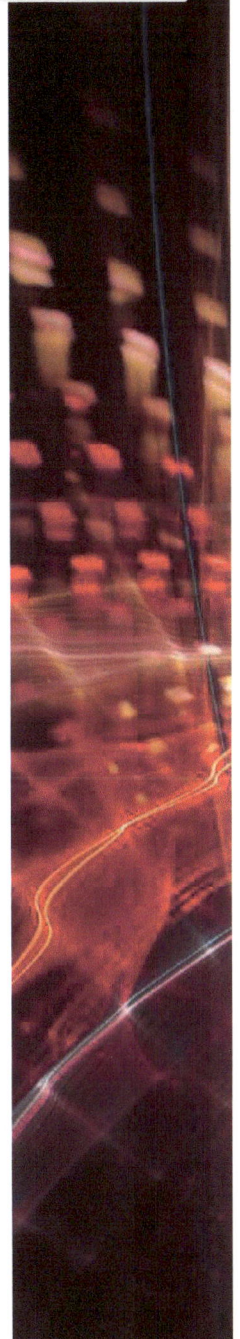

20. Give an example of a situation where you would apply the principles of engineering economics?

21. Describe an engineering project you have worked on and explain its importance to society in 30-40 words.

22. Describe 1-3 parameters that are most important to heat transfer.

23. Describe 1-3 parameters that are most important to thermodynamics.

24. Describe 1-3 parameters that are most important to fluid mechanics.

25. Describe safety concerns to boil water or an unknown liquid.

26. Describe a few jobs a mechanical engineer can do in a nuclear power plant.

27. Describe a few jobs a mechanical engineer can do in a solar thermal power plant.

28. List at least 8 separate hand-held tools useful for assembling, disassembling, and repairing mechanical or electrical devices (different attachments, sizes, and shapes do not count as separate tools).

29. Define product validation and product verification. Explain how safety considerations are involved in each.

REFERENCES

[1] Y. Cangel and M. A. Boles, "Thermodynamics: An Engineering Approach 4th Edition in SI Units," Singapore (SI): McGraw-Hill, 2002.

[2] F. P. Incropera, A. S. Lavine, T. L. Bergman, and D. P. DeWitt, Fundamentals of heat and mass transfer. Wiley, 2007.

[3] I. H. Shames and I. H. Shames, Mechanics of fluids. McGraw-Hill New York, 1982.

[4] R. S. Figliola and D. E. Beasley, "Theory and design for mechanical measurements," ed: IOP Publishing, 2001.

[5] HF Lecture Bottle Explosion. Available: https://www.ehs .ucsb.edu/files/docs/ls/HF_lecturebottle.pdf

[6] Accident Report: Rupture of Vacuum Flask. Available: https://www.ehs.ucsb.edu/files/docs/ls/VacuumFlask.pdf

[7] Chemistry Lab Fire at University of Texas. Available: https://www.ehs.ucsb.edu/files/docs/ls/UT_fire.pdf

[8] Chemical Stored on Floor Causes Fire. Available: https://www.ehs.ucsb.edu/files/docs/ls/UCSC_fire.pdf

[9] "Independent Accident Investigation," 2001, Available: https://www.ehs.ucsb.edu/files/docs/ls/UCI_fire.pdf.

[10] Fighting Lab Fires. Available: https://www.ehs.ucsb.edu/files/docs/ls/Ohio_fire.pdf

[11] UCLA Lab Fire Fatality. Available: https://www.ehs.ucsb.edu/files/docs/ls/UCLA.accident.summary.pdf

[12] Michele Dufault '11 dies in Sterling Chemistry Laboratory accident. Available: https://yaledailynews.com/blog/2011/04/13/michele-dufault-11-dies-in-sterling-chemistry-laboratory-accident/

[13] Anaerobic Chamber Explosion. Available: https://ehrs.upenn.edu/healthsafety/lab-safety/safety-alerts/anaerobic-chamber-explosion

[14] Report to the University of Hawaii at Manoa on the Hydrogen/Oxygen Explosion of March 16, 2016. Available: https://cls.ucla.edu/images/document/Report%201%20UH.pdf

[15] Lab Experiment Goes Wrong, Starts Fire at U-Md. Available: https://www.washingtonpost.com/local/public-safety/lab-experiment-goes-wrong-starts-fire-at-u-md/2019/05/01/cf266f44-6c22-11e9-a66d-a82d3f3d96d5_story.html

[16] PPE Signs. Available: https://www.safetysign.com/

[17] Fire Safety Online Training Available: https://www.fau.edu/divdept/envhs/Fire-exting/firesafetytraining.pdf

[18] Compressed Gas Cylinder Training. Available: https://www.pfw.edu/offices/rem/safety-training/doc/Compressed %20Gas%20Cylinder%20Training.pdf

[19] High Voltage Qualified. Available: https://www.e-hazard.com/arc-flash-training/pdf/high-voltage-training-outline.pdf

[20] Biosafety Training Program. Available: https://som.uci.edu/bsl3-training/about.asp

[21] Chemical & Lab Safety Training. Available: https://sc.edu/about/offices_and_divisions/ehs/training/research_laboratory_safety/chemical_and_lab_safety_training/index.php

[22] Radiation Safety Training. Available: https://www.ehs.uci.edu/programs/radiation/radtrain.html

[23] Laboratory Safety – Beyond The Fundamentals Workshop Description. Available: http://dchas.org/2018/04/ 13/beyond-the-fundamentals/

[24] Hazard Communication Standard Pictogram. Available: www.osha.gov

[25] The BRUCE Zone. Available: http://thebrucezone.weebly.com/lab-safety.html

ANALYSIS OF EXPERIMENTAL DATA

2.1. INTRODUCTION

In an experiment, one major task is to conduct measurements for collecting data. A large number of measurement data can be the direct outcome of experimental work. In general, the more data the better. Statistics is a popular valuable tool for experimentalists to conduct data analysis and eventually draw conclusions from data processing. The mean of a data sample is usually used as the final result for a measurement. The standard deviation of the sample measures the confidence of the final result and is usually used as additional information. In addition, each measurement needs to be independent so that the data set is equally weighed. Statistics may be used in experimental design and plan. Indeed, before conducting an experiment, several aspects need to be considered in preparation, including the selection of apparatus, relevant mathematical correlations, and uncertainty estimate of the final results. Selecting the proper apparatus is essential to any experimental work. For example, temperature measurement requires thermometers. There are several types of thermometers with various ranges and resolutions. A high-resolution apparatus is usually expensive and requires training before use. However, low-resolution apparatuses usually lead to a large standard deviation or uncertainty in the final result. In engineering applications, uncertainty needs to be within tolerance to avoid component mismatch or design failures. Understanding how the apparatus' resolution is related to the measurement error or uncertainty and how the error or uncertainty propagates to the final value is thus fundamentally important for experimental design, which will be introduced in this chapter.

2.2. ERROR AND UNCERTAINTY ANALYSIS

Prior to the experiment, it is important to estimate the effect of measurement errors in final results. For example, there is an error or uncertainty in the measurement of time using a stopwatch. Then, using this stopwatch in the speed measurement of a car in a freeway, the error or uncertainty in the time measurement will then influence the final result of the car speed. This is practically important for issuing speed tickets on the road. To evaluate the error or uncertainty in a final result, one needs to:

(1) describe the relationship between the final result and a measured quantity, and

(2) estimate the error or uncertainty in the measurement of the quantity.

A simple example is the temperature measurement from a thermocouple, in which temperature is the final result or target and the voltage is the measured quantity. In other words, we directly measure the voltage in a thermocouple and then use a formula to convert the voltage to temperature. Assuming the relationship between temperature and voltage is linear with a calibration constant K:

$$T = Ke \qquad\qquad\qquad 1$$

The voltage error, δe, in the measurement will then lead to an error in the final temperature:

$$\delta T = K\delta e \qquad\qquad\qquad 2$$

Since the calibration equation is linear, the temperature error is independent of the magnitude temperature itself. For this reason, linear sensors are often desirable. Another example is the pressure transducer, which converts the voltage signal to pressure using a linear relationship.

In contrast to thermocouples, the flow rate measurement using the data of volume and time measurement is non-linear, involving multiple variables, *i.e.* the flow rate $\dot{Q}$ is the volume V divided by the elapsed time t that the flow rate takes to fill the volume V, as below:

$$\dot{Q} = \frac{V}{t} \qquad\qquad\qquad 3$$

Mathematically, the error in $\dot{Q}$, *i.e.* $\delta\dot{Q}$, is obtained by applying the chain rule of differentiation to Equation 3:

$$\delta\dot{Q} = \frac{1}{t}\delta V - \frac{V}{t^2}\delta t \qquad\qquad\qquad 4$$

where δV and δt are errors in the volume and time measurements, respectively. In general, errors fall into two categories:

Bias and Uncertainty.

Bias is an average difference between the measurement from a device and that from an accepted standard. For example, a watch might be biased by two minutes ahead of the accurate time from an accepted reference such as the atomic clock at the US Naval Observatory. On a weight scale, you need to reset it in order to have an accurate evaluation of your body weight. Uncertainty is the error due to reading the measurement (*e.g.* estimate below the smallest scale division) and/or fluctuations inherent in an equipment (*e.g.* noises). For noises, a digital weight scale may give a fluctuating number in the display, which may be due to the noise from the pressure sensor or voltage reading. Note that the body weight fluctuation from morning to night may be treated as a 'noise' in your body, as it is not associated with the scale. In general, uncertainty is precision plus noises and accuracy is bias plus uncertainty. Note that a measurement can have high precision but low accuracy; high accuracy does require high precision. We exclude gross errors or blunders, such as reading the wrong scale or transcribing the data and omitting a factor of 10, wrong units, *etc*.

To show the mathematical relationship, the error in a variable X, *i.e.* δX, is the sum of the bias and random parts:

$$\delta X = \epsilon_X + \delta x \qquad\qquad\qquad 5$$

where ϵ_X is the bias and δx is the uncertainty or random part such that $\|\delta x\| = 0$. The former is an average over many readings or the time average of a fluctuating reading. In the following discussion, the biases are neglected and assumed to be zero after calibration. As to uncertainty, we assume it arises from estimate below the smallest scale division only. Under usual conditions, our human eyes can fairly well estimate whether a reading is above or below the middle of the smallest scale division. Conventionally, the uncertainty is set to be:

Half of the Smallest Scale Division or Precision.

Note that in practice, the measurement uncertainty is dependent on experience.

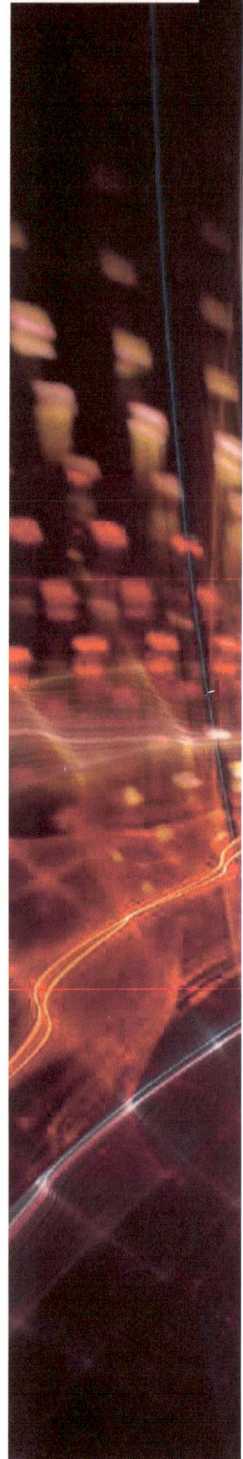

#Example 1: For a ruler, stopwatch, and voltage meter of 1 mm, 0.01 s and 0.001 V precisions, respectively, what are their experimental uncertainties associated with the precisions? If the voltage meter is to measure a thermocouple voltage, what will be the uncertainty in the final temperature reading if K=200 °C/V in Equation 2?

Solution:

$$\delta x = 0.5 \ mm, \quad \delta t = 0.005 \ s, \ and \ \delta e = 0.0005 \ V.$$

Apply Equation 2,

$$\delta T = K \delta e = 0.1 \ ^{\circ}C$$

In the flow rate measurement, because the volume and time are obtained by two apparatuses, the errors in these two variables are independent. As a result, the *root-mean-square (rms) method* applies from Equation 4:

$$\sqrt{\overline{(\delta \dot{Q})^2}} = \sqrt{\frac{1}{t^2}\overline{(\delta V)^2} + \frac{V^2}{t^4}\overline{(\delta t)^2}} \longrightarrow \quad 6$$

In the above, we assume that $\overline{\delta V \cdot \delta t} = 0$. In other words, the average of two random errors is uncorrelated. To understand why the *rms* form is taken, one can treat the right side of Equation 4 as a complex number (a + bj), with the first term in Equation 4 representing the real part (a) and the second the imaginary part (b). For t and V, if multiple measurements are taken, one can use their averages. In general, the person who takes the measurement should have an idea of the range of t and V in experiment.

#Example 2: for δV = 0.5 ml, δt = 0.2 sec, V = 0.955 liter and t = 10.8 sec, calculate the rms error.

Solution:

$$\sqrt{(\delta\dot{Q})^2} = \sqrt{2.14\cdot 10^{-9} + 2.68\cdot 10^{-6}} = 0.001638 \; l/sec \qquad \textbf{7}$$

The flow rate can be written as

$$\dot{Q} = 0.088\,40 \pm 0.01638 \; l/sec \; or \; \dot{Q} = 0.0884 \; l/sec \pm 1.9\% \qquad \textbf{8}$$

where the former part 0.0884 comes from the calculation of V/t directly.

The full ranges also impact the rms errors. One can repeat the above calculation by using V = 95.5 liter and t = 10.8 sec.

Note that when the average of a number of trials is reported, it is acceptable to report the answer to one more significant figure than in the individual trial, *e.g.*, the average of 2 and 3 is 2.5. Conversion factors and known constants, *e.g.*, π do not affect the number of significant figures.

2.3. PROBABILITY DISTRIBUTION

2.3.1. Gaussian Distribution

The Gaussian distribution, also called the normal distribution, is the most common distribution used in statistics. The Gaussian distribution function is given by:

$$P(x) = \frac{1}{\sigma\sqrt{2\pi}} \exp\left(-\frac{(x-\mu)^2}{2\sigma^2}\right) \qquad \textbf{9}$$

where the two parameters μ and σ^2 are corresponding to the mean and variance of the distribution. In experimental measurement, errors or uncertainties arising from random events can usually be described by the Gaussian distribution.

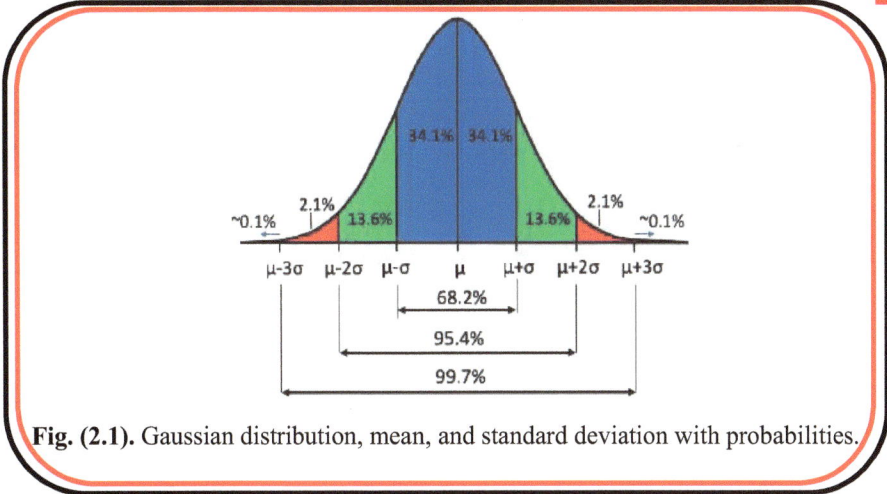

Fig. (2.1). Gaussian distribution, mean, and standard deviation with probabilities.

The shape of the Gaussian distribution is shown in Fig. (2.1), which is often called a "bell" curve. The physical meaning of the mean and the standard deviation is clearly shown: the former is at the mean value of the curve while the latter corresponds to the half-width of the peak at about 60% of the full height.

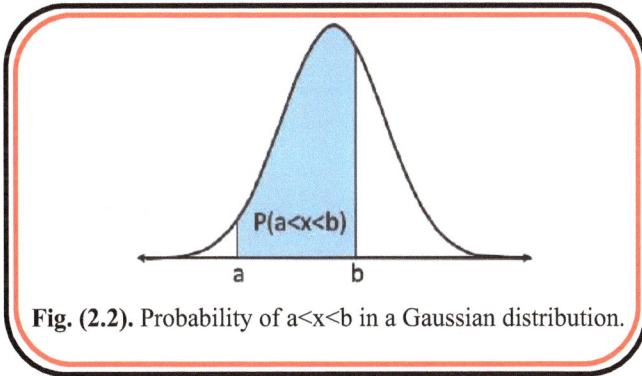

Fig. (2.2). Probability of a<x<b in a Gaussian distribution.

The value of the Gaussian curve or distribution function does not represent the probability of an event at x. Instead, the integral of the function from x=a to b, $\int_a^b P(x)\,dx$, *i.e.* the area bounded by the curve, x-axis, and the two vertical lines at x=a and x=b, does, as shown in Fig. (2.2). The integral, unfortunately, cannot be calculated analytically; thus, one needs numerical integration. Tables of integral values are readily found and can be developed in terms of a reduced Gaussian distribution under $\mu = 0$ and $\sigma^2 = 1$. All the Gaussian distributions can be unified to this reduced form (see Fig. 2.3) through the below transformation:

$$z = \frac{x-\mu}{\sigma} \qquad\qquad\qquad\qquad \textbf{10}$$

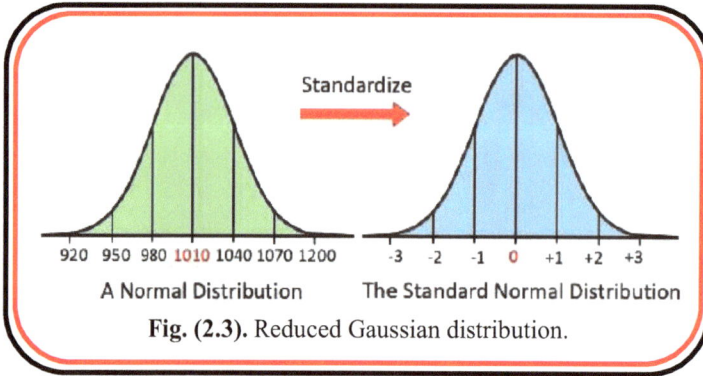

| 920 950 980 **1010** 1040 1070 1200 | -3 -2 -1 **0** +1 +2 +3 |
| A Normal Distribution | The Standard Normal Distribution |

Fig. (2.3). Reduced Gaussian distribution.

In addition to the Gaussian distribution, other probability distributions exist. For example, the Rayleigh and Weibull probability distribution function (PDFs) are frequently used to describe the wind speed (U) distribution over time, as shown in Fig. (**2.4**).

The Rayleigh PDF and cumulative distribution function, C(U), are given by:

$$P(U) = \frac{\pi}{2}\left(\frac{U}{\bar{U}^2}\right) exp\left[-\frac{\pi}{4}\left(\frac{U}{\bar{U}}\right)^2\right] \qquad\qquad \textbf{11}$$

$$C(U) = \int_0^U P(U)\,dU = 1 - exp\left[-\frac{\pi}{4}\left(\frac{U}{\bar{U}}\right)^2\right] \qquad\qquad \textbf{12}$$

The Weibull PDF and cumulative distribution function are given by:

$$P(U) = \left(\frac{k}{c}\right)\left(\frac{U}{c}\right)^{k-1} exp\left[-\left(\frac{U}{c}\right)^k\right] \qquad\qquad \textbf{13}$$

$$C(U) = 1 - exp\left[-\left(\frac{U}{c}\right)^k\right] \qquad\qquad \textbf{14}$$

Mean wind speed: $\bar{U} = \int_0^\infty UP(U)dU$ $\qquad\qquad$ **15**

Standard deviation: $\sigma = \sqrt{\int_0^\infty (U-\bar{U})^2 P(U)dU}$ $\qquad$ **16**

Both k and c are functions of the mean wind speed and the standard deviation of the wind speed. The Rayleigh PDF only depends on one parameter, *i.e.* the mean wind speed, making it much easier to use. Note that the Rayleigh distribution is a special case of the Weibull distribution by setting $k = 2$ and $c = U$. Table **2.1** lists popular probability distribution functions with their shapes.

Table 2.1. Probability distribution function.

Type	Application	Shape
Normal/Gaussian Distribution	A type of continuous probability distribution for a real-valued random variable. The most important probability distribution because it fits many phenomena.	
Bernoulli Distribution	A discrete distribution of a random variable which takes a binary, boolean output: 1 with probability p, and 0 with (1-p). In clinical trials, it can describe a single individual experiencing an event like death, a disease, or disease exposure.	
Uniform Distribution	A probability distribution that has constant probability. An important application is the generation of random numbers.	
Binomial Distribution	The probability of a success or failure outcome in an experiment or survey that is repeated multiple times. It can analyze repeated independent trials, especially the probability of meeting a threshold given an error rate, and thus is suitable to risk management.	
Rayleigh Distribution	A continuous probability distribution for nonnegative-valued random variables. Widely used in life testing experiments, wind speed probability distribution, reliability analysis, applied statistics and clinical studies.	
Weibull Distribution	A continuous and tri-parametric probability distribution. Widely used in reliability and life data analysis and wind speed probability distribution.	
Poisson Distribution	The probability of a given number of events occurring in a fixed interval of time or space if occurring at a constant mean rate and independently of the time since the last event. It can also be used for the number of events in other specified intervals such as distance, area, or volume.	

(Table 2.1) cont.....

Exponential Distribution	The probability of the time between events in a Poisson point process. It can be used in a range of disciplines including queuing theory, physics, reliability theory, and hydrology.	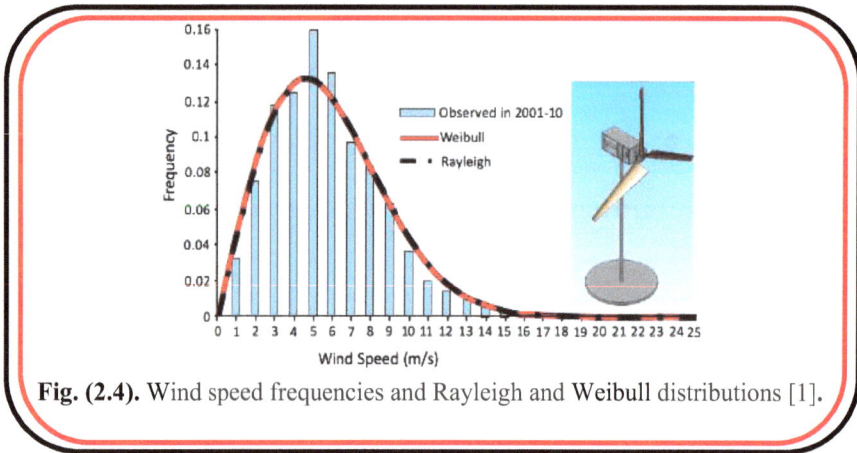

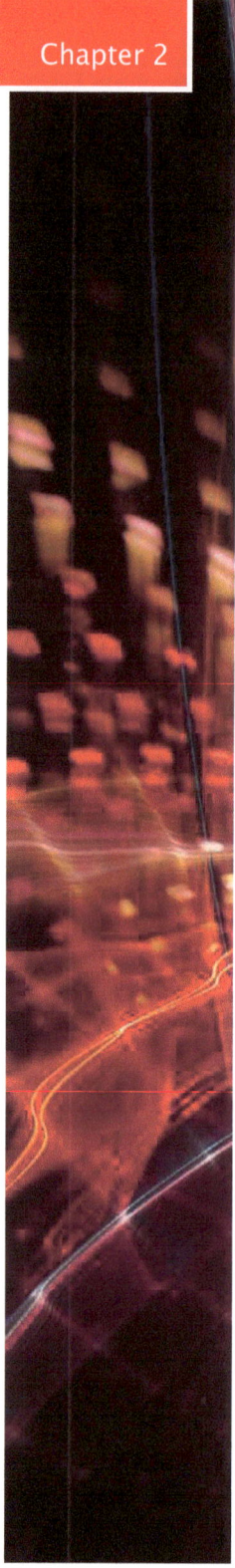

Fig. (2.4). Wind speed frequencies and Rayleigh and Weibull distributions [1].

2.3.2. Average and Mean Value

The average or mean value is the sum of the number of readings divided by the number of samples. However, implicit in that definition is that we are obtaining the 'population mean'- we are averaging over all possible samples. In the experiment, this is not possible and we take a finite subset of the population and obtain a 'sample mean'. Formally, we call the population mean μ and the sample mean $\bar{x}$:

$$\mu = \lim_{N \to \infty} \frac{1}{N} \sum_{i=1}^{N} x_i \qquad\qquad 17$$

and

$$\bar{x} = \frac{1}{N} \sum_{i=1}^{N} x_i \qquad\qquad 18$$

where N means the number of samples and x_i is the sample value. Note that the above formula applies to a data sample following the Gaussian distribution. For other distributions, the probability distribution function needs to add as the weight, as shown in Equation 15.

2.3.3. Standard Deviation

Before proceeding with the formulas for standard deviation, *i.e.* the square root of the variance, we need to address a subtle point. Since we are concerned with the deviation about the mean, we need to know the mean. There are two options for this: 1.) we can know the population mean, μ, independently. 2.) we have to get an estimate of the population mean from a sample that we are also using to get the variance. In the latter case, we run through the data twice: first to get $\bar{x}$; second, to get the sample standard deviation. It turns out that while $\bar{x}$ is an unbiased estimator of μ, the quantity

$$\frac{1}{N}\sum_{i=1}^{N}(x_i - \bar{x})^2 \qquad\qquad\qquad \text{19}$$

is not. It is a biased estimator of the population variance. The reason is that the number of samples is viewed as the degrees of freedom in the calculation, and we have used one of these to estimate the population mean μ with $\bar{x}$. Therefore, the correct unbiased sample variance is given by the square of the standard deviation s_x:

$$s_x{}^2 = \frac{1}{N-1}\sum_{i=1}^{N}(x_i - \bar{x})^2 \qquad\qquad \text{20}$$

while the variance of the population with a pre-given mean μ is expressed by:

$$\sigma_x{}^2 = \frac{1}{N}\sum_{i=1}^{N}(x_i - \mu)^2 \qquad\qquad \text{21}$$

An easy way to remember which formula to use is to assume you have a sample of one data point in the experiment, *i.e.* N=1. If the mean μ is a given value independent of the measurement, the deviation will be calculated using the variance formula, Equation 21. If the mean μ is not given and is calculated by this sample data, *i.e.* equal to the value of the data point, your deviation is uncertain as it is unclear your experiment data is reliable.

Then the unbiased sample variance of Equation 20 is suitable, which gives 0 divided by 0. By using the variance formula of Equation 21, one will get 0 divided by 1, equal to 0, which is meaningless. The point is that when you use software or a calculator, make sure which formula is being used. Obviously, the difference is negligible when N is sufficiently large.

#Example 3: for a data set of x(i): (3.0, 3.5), (a) calculate Equations 19 and 20. (b) If the mean μ is known to be 3.25, please calculate Equations 20 and 21.

Solution:

(a) From Equation 18,

$$\bar{x} = 3.25$$

Apply Equations 19 and 20, one will get, respectively:

$$0.0625 \text{ and } 0.125$$

Note that the two formulas will lead to very different results for a small sample size.

(b) For μ is known to be 3.25, the sample mean is the same as that of Equation 18.

Apply Equations 20 and 21, one will get, respectively:

$$0.125 \text{ and } 0.0625.$$

Note that if the mean is pre-given, Equation 19 should be used for the sample variance.

Again, the above formula applies to samples following the Gaussian distribution. For other distributions, the formula needs revision by using the distribution function as the weight, as shown in Equation 16 for an example.

2.3.4 Frequency Distribution

Frequency distribution provides information on the number of times a value or event occurs, and thus it follows a similar profile as the probability density distribution. The Gaussian distribution represents a frequency distribution in which the peak frequency occurs at the mean value, and events are symmetrically distributed at the two sides of the mean, as shown in Fig. (**2.5**) as an example.

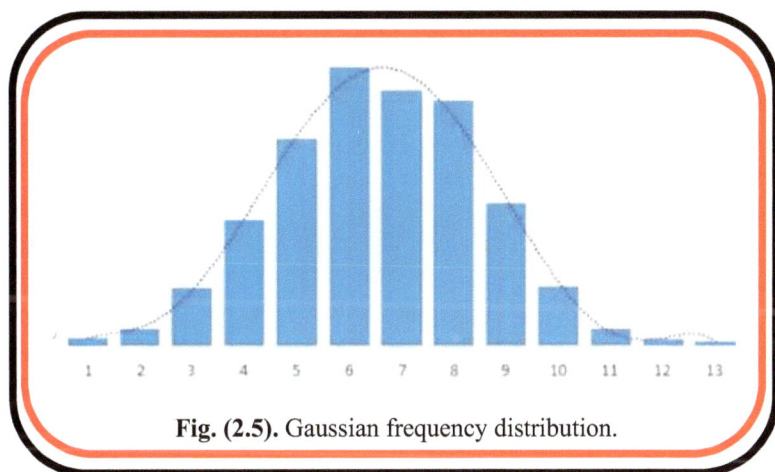

Fig. (2.5). Gaussian frequency distribution.

The calculation of the frequency distribution – sometimes called the Histogram – is easy in Matlab (hist), but not in Excel. In Excel, one needs to set up the vector (column) of bins for the data sorted into. Then, an array operation follows **Frequency** under **Statistics** in f_x, which places the counts for each bin adjacent to the bin column. The counts represent the frequency of a number falling in the bin. The choice of the bins for the frequency distribution is a choice to be made by the individual. A few large bins give a smooth distribution but no detail; many bins show excessive detail, but the underlying distribution cannot be ascertained. A balance should be struck. There is no simple formula; it is a value judgment. Obviously, extremes such as one or two bins or the number of bins greater than the number of samples should be avoided. Attempts can be made to test several numbers and sizes of bins to find the best options.

It is also the choice of experimentalists to interpret the frequency distribution to follow specific distributions on the basis of physics and common sense. Not all data should follow a "bell" curve shape. An example is given in the wind speed distribution, as shown in Fig. (**2.4**).

2.4. OUTLIERS AND THREE SIGMA TESTING

In experimental data, it is not uncommon to find a data point or a few that differ significantly from observation in expect. Such data points, also called outliers, can cause serious problems in statistical analyses. Outliers may result from variability in measurement, abnormal action in data collection, typos or errors in data recording, or even mistyping of units. In most cases, outliers should be excluded from the data set in analyzing the final result. A few approaches have been proposed for outlier detection, such as Chauvenet's criterion and three-sigma test. The former identifies outliers if they have less than a 1/2N probability of occurrence, where N is the sample size. The latter detects outliers that lie outside the range of 99.7% probability of occurrence, as shown in Fig. (2.1). The major steps in a three-sigma test on a data set x(i) include:

> **Step I:** Calculate the mean $\bar{x}$ and standard deviation s_x of x(i): i=1 to N;
> **Step II:** For each data point j (j=1 to N), if $|x(j)-\bar{x}|>3\underline{s}_x$, then, x(j) is an outlier; the number of the outliers is labeled by Nj;
> **Step III:** Eliminate all the outliers in x(i), which leads to a new data set x(i), where i is renumbered from 1 to N-Nj;
> **Step IV:** if Nj >0, go to Step I for the new data set x(i): i=1 to N-Nj. Otherwise, there are no outliers in x(i).

Note that the method assumes the data follow a Gaussian distribution, which may not be true for some cases, *e.g.* the Rayleigh distribution. Table **2.2** lists several popular outlier detection methods.

In addition, the outliers are clearly "visible" in the frequency distribution, as shown in Fig. (**2.6**). Thus, outliers need to be eliminated in order to obtain a bell shape.

#Example 4: for a data set of x(i): (1.00, 1.00, 0.99, 0.98, 1.00, 1.02. 1.01, 1.00, 10.00, 1.00, 1.01, 0.99, 0.98, 0.90, 0.96, 1.05), please calculate $\bar{x}$ and s_x. Do you find any $|x(j)-\bar{x}|>3s_x$? Now get rid of any outliers. Without calculation, can you estimate the sample mean and compare it with the one with the outliers.

Solution:

Apply Equations 18 and 20,

$\bar{x} = 1.56$ and $s_x = 2.25$

Among the data points, $x(j)=10.00$ will lead to :

$$|10.00-\bar{x}|>3s_x$$

Thus, it is an outlier. For the new data set: (1.00, 1.00, 0.99, 0.98, 1.00, 1.02. 1.01, 1.00, 1.00, 1.01, 0.99, 0.98, 0.90, 0.96, 1.05). The new sample mean is estimated to be around 0.99-1.00, much less than the original 1.59.

Thus, the outliers need to be removed for the final calculation of the sample mean.

Note that there may be additional outliers in the new data set: (1.00, 1.00, 0.99, 0.98, 1.00, 1.02. 1.01, 1.00, 1.00, 1.01, 0.99, 0.98, 0.90, 0.96, 1.05).

Table 2.2. Outlier detection methods.

Method	Description
3 Sigma Test [2]	A simple widely used method that outliers are outside of three standard deviations from a mean.
Chauvenet's Criterion [3]	It compares the maximum allowable deviation to the difference between a suspected outlier and the mean divided by the standard deviation.
Z-Score [4]	Also called a standard score, it gives an idea of how far from the mean a data point is. It measures how many standard deviations from the mean.
DBSCAN (Density Based Spatial Clustering of Applications with Noise) [5]	A density based clustering algorithm is focused on finding neighbors by density (MinPts) on an 'n-dimensional sphere' with radius ε.
Isolation Forests [6]	An effective method for detecting outliers or novelties in data, which is based on binary decision trees.
Gaussian Regression [7, 8]	Data points with a probability lower than a particular threshold are treated as outliers.
Poisson Regression [9]	Outliers are observations that do not follow the Poisson distribution of the majority of the data.

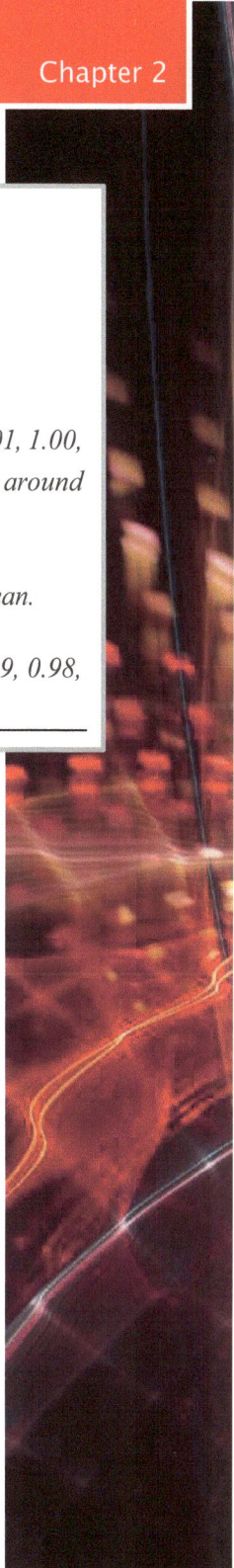

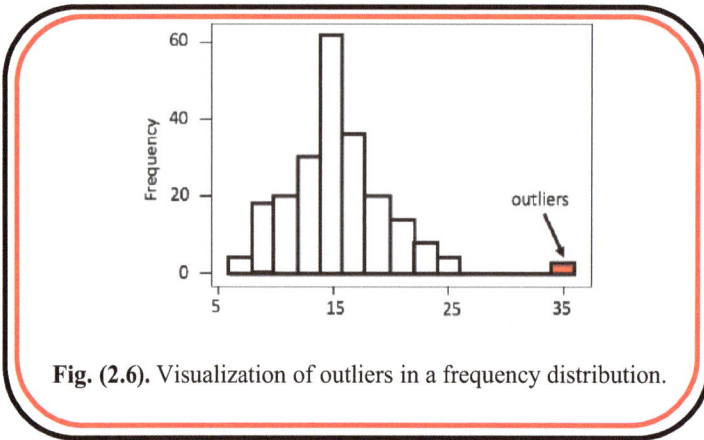

Fig. (2.6). Visualization of outliers in a frequency distribution.

2.5. EXPERIMENT

As an example to examine the statistical nature of experimental data and error analysis, an experiment is introduced below to conduct a number of independent measurements of a water flow rate. In the experiment, the flow rate is adjusted by the pipe outlet height. At a fixed flow rate, a number of two measurements will be taken, including the volume and the duration that it takes for the flow rate to fill the volume. The direct measurements of volume and time have errors or uncertainties associated with equipment. This experimental setup is a popular way for flow meter calibration, introduced in Chapter 5 in detail.

2.5.1. Apparatus

The flow bench is set up to control a pressure drop between the pipe inlet and outlet, adjusted by the vertical outlet position. In pipe flows, a pressure drop will drive a corresponding flow rate. To measure a flow rate, two basic instruments are used, including a graduated cylinder for the volume measurement and a stopwatch for time. Note that any bias in the stopwatch needs to be corrected if the hand does not reset to zero. Before taking measurements, one can estimate the precision of each instrument based on the divisions of the scales and how close one reliably reads them.

2.5.2 Procedure

a.) Set the pipe outlet to a position for a constant flow rate. An appropriate flow rate should be set to fully utilize the precisions and full scales of the graduated cylinder and stop watch. For example, it should take about 5-10 seconds to fully fill the graduated cylinder.

b.) Measure the water volume and time using the graduated cylinder and stopwatch, respectively. Let the graduated cylinder almost fill to its full capacity to minimize errors.

c.) Alternate other persons to repeat the measurements. In this way, the set of the obtained data points are independent. For statistical analysis, it is recommended to conduct at least 35 independent measurements for each flow rate.

d.) After obtaining all the data for one flow rate, set the pipe outlet to another position for a different flow rate, repeat a)-c) steps.

A sample of time and volume data is given in Table **2.3** for three cases, including two different flow rates and two different size graduated cylinders. In an experiment, the precisions of the time and volume measurements as well as other relevant information, should be written on a datasheet attached to the measurement data. In most cases, the precisions can be estimated from the last digit of the data number.

#Example 5: without prior knowledge of the experimental equipment's precisions, estimate precisions and uncertainties in the Small Cylinder data in Table 2.3.

Solution:

The last digit in the volume is at the scale of 0.1 ml, which should be guessed from the reading. Thus, we can estimate the precision is a scale above, i.e. 1 ml, and the uncertainty is 0.5 ml.

As to the stopwatch, following the same strategy, the precision is estimated to be 0.1 s and the uncertainty is 0.05 s.

Note that the uncertainty, in general, indicates an estimate for evaluation purposes. The precision for the graduated cylinder can be 0.12 ml, though a rare case, with an uncertainty of 0.6 ml. For evaluation purpose, there is no major difference between 0.5 and 0.6 ml.

Table 2.3. Sample of experimental data.

Sample#	Large Cylinder (fast)		Large Cylinder (slow)		Small Cylinder (fast)	
	volume(ml)	time (sec)	volume(ml)	time (sec)	volume(ml)	time (sec)
1	684	5.11	825	13.30	26.8	0.63
2	731	5.30	1000	16.43	36.9	0.65
3	681	4.88	630	10.36	35.1	0.79
4	811	5.98	670	11.06	41.2	0.55
5	840	6.11	660	10.69	43.5	0.66
6	730	5.32	620	10.28	42.2	0.68
7	685	5.00	610	10.12	37.9	0.72
8	701	5.04	710	11.50	38.8	0.66
9	660	4.87	560	8.98	47.1	0.60
10	700	5.11	625	10.39	47.2	0.56
11	697	5.06	640	10.69	39.5	0.65
12	679	4.90	660	11.03	34.8	0.40
13	665	4.92	643	10.45	35.5	0.55
14	630	4.52	653	10.71	41.5	0.65
15	660	4.93	703	11.75	48.8	0.60
16	671	5.05	615	10.33	47.8	0.59
17	730	5.36	559	9.33	41.3	0.61
18	720	5.42	681	11.43	43.9	0.68
19	780	5.58	659	10.98	41.4	0.66
20	680	5.11	625	10.28	36.2	0.53
21	730	5.40	651	10.65	35.8	0.50
22	670	5.01	644	10.65	37.1	0.55
23	671	5.06	652	10.64	41.0	0.55
24	670	5.06	649	10.68	27.9	0.68
25	660	4.86	652	10.77	43.9	0.55
26	650	4.89	648	10.75	40.1	0.52
27	660	4.87	548	9.17	41.8	0.59
28	710	5.13	615	10.30	35.8	0.69
29	610	4.67	640	9.90	46.1	0.62
30	611	4.53	695	10.73	48.7	0.66
31	680	5.05	601	10.03	26.1	0.46
32	640	4.77	612	10.11	44.2	0.42
33	690	5.24	671	11.27	39.9	0.46
34	650	4.82	598	10.12	34.9	0.41
35	670	4.92	652	10.90	33.8	0.45
36	660	5.01	656	11.15	41.4	0.43
37	661	4.92	605	10.23	36.1	0.35
38	663	4.79	731	12.23	31.2	0.46
39	685	5.20	619	10.38	29.5	0.50
40	665	4.96	665	11.28	43.1	0.42
41	660	4.99	620	10.33	43.8	0.53
42	721	5.32	635	10.56	40.7	0.72
43	730	5.47	620	10.37	33.2	0.36
44	640	4.72	625	10.39	45.9	0.50
45	675	5.05	615	10.31	40.7	0.39
46	670	4.89	705	11.69	37.6	0.43
47	670	4.87	597	10.03	41.6	0.55
48	720	5.12	620	10.37	36.5	0.39
49	701	5.07	604	10.05	49.1	0.43
50	746	5.50	660	10.97	43.5	0.40

2.5.3. Error Analysis

In practice, this can be done prior to conducting the experiment. As an example to show how the error or uncertainty theory is related to the standard deviation and distribution of experimental data, one can estimate the error in the final flow rate using Equation 6 and the values of V, t, δV and δt in the experiment. V and t can use the volume and time averages in the experiment, while δV and δt can use half of the precisions of the graduated cylinder and stopwatch, respectively. It can be seen from Equation 6 that a large δV or δt will lead to a bigger $\delta\dot{Q}$. In the distribution of the experimental $\dot{Q}$, the standard deviation may positively correlate with $\delta\dot{Q}$.

2.5.4 Statistics and Distribution of Experimental Data

Using the set of V_i and t_i measurement data, one can calculate $\dot{Q}_i$ and conduct statistic and distribution analysis, including

    a. Mean
    b. Standard Deviation $s_{\dot{Q}}$
    c. Frequency Distribution

The analysis can be done using: 1.) hand calculation or calculators based on the definition; 2.) the Excel "function wizard, fx" and chose Statistics; or 3.) Matlab or another software. In practice, the sample means or average after eliminating the outliers is used as the final result of the experimental flow rate. The standard deviation measures the degree that the measurement may deviate. In commercial flow meters, the manufacturing data sheet should list the full range and precision. The standard deviation $s_{\dot{Q}}$ and error estimate $\delta\dot{Q}$ should positively correlate to the flow meter precision as given in the data sheet. Note that the standard deviation and uncertainty should not be the same because of their independent definitions. The frequency distribution, if following the Gaussian distribution, visually shows the mean, standard deviation, and presence of outliers. In calculating the frequency distribution of $\dot{Q}$, the N values of $\dot{Q}$ can be sorted into m bins of size $\Delta\dot{Q}$ spanning the range $\dot{Q}_{min}$ to $\dot{Q}_{max}$. Note that the larger the N, the better profile one can achieve. However, this may not be true for the selection of m. In an extreme case, m is sufficiently large that each bin only has one data at the most. The resulting distribution profile is meaningless. It requires experience and several trials to achieve the best distribution profile. In general, for 35-50 data points, 6-10 bins may work the best.

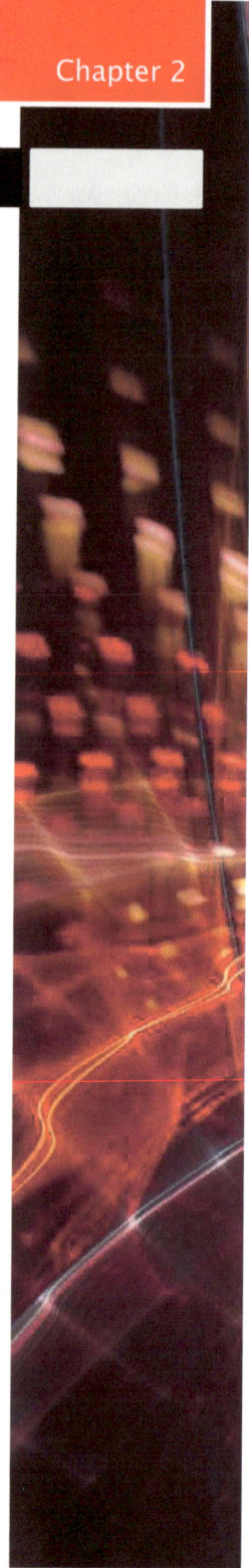

2.6. QUESTIONS

1. In the sample of experimental data, please estimate the precisions and full range of each equipment.

2. Plot the frequency distributions of $\dot{Q}$. Comment on the distributions: are there any outliers? Please get rid of the outliers. Why are the distributions different?

3. Evaluate Equation 6 using the average values of t and V (without outliers) to determine the predicted root-mean-square error $\delta \dot{Q}$.

4. Compare the root-mean-square error $\delta \dot{Q}$ with the square root of the correct unbiased sample variance, *i.e.* standard deviation s_Q, and comment.

5. Plot the reduced Gaussian distributions for three deviations of 1, 2 and 3, respectively, using Matlab or other programming languages, and attach the codes.

6. Plot the Rayleigh distributions for three means 1 m/s, 5 m/s, and 10 m/s, respectively, using Matlab or other programming languages, and attach the codes.

7. Give an example, sketch in a plot, to show what an outlier looks like in a Rayleigh distribution. Can we use 3 sigmas testing to find it? Explain.

8. What are the two categories of experimental errors? Please give examples using a ruler and stopwatch.

9. Does high precision always lead to high accuracy? Why or why not?

10. For the given data sample, please set the number of the bin to 50 and 200 and plot the distribution profiles.

11. In a thermocouple, voltage (e) is measured to be 1 V. K= 300 °C/V. $\delta e = 0.001$ V, then $\delta T = ?$

12. In Example 2, redo the calculation by using V = 0.0955 liter and t = 1.08 sec.

13. In Example 3, add the extra data (3.1, 3.4, 3.2, 3.3) and redo the calculations.

14. In Example 4, plot the frequency distribution and visually point out any outliers.

15. In the new data set of Example 4, please do one more round of 3 sigma testing for additional outliers and the sample mean.

16. List three examples that follow Gaussian distribution.

17. List 1-3 examples that do NOT follow Gaussian distribution.

18. Calculate the mean and standard deviation in the ages of your family members.

REFERENCES

[1] S. H. Pishgar-Komleh, A. Keyhani, and P. Sefeedpari, "Wind speed and power density analysis based on Weibull and Rayleigh distributions (a case study: Firouzkooh county of Iran)," (in English), Renewable & Sustainable Energy Reviews, vol. 42, pp. 313-322, 2015.

[2] R. Lehmann, "3 sigma-Rule for Outlier Detection from the Viewpoint of Geodetic Adjustment," (in English), Journal of Surveying Engineering, vol. 139, no. 4, pp. 157-165, 2013.

[3] Chauvenet's Criterion. Available: https://en.wikipedia.org/wiki/Chauvenet%27s_criterion

[4] Statistics How To. Available: https://www.statisticshowto.datascience central.com/probability-and-statistics/z-score/

[5] M. Ester, H.-P. Kriegel, J. Sander, and X. Xu, "A density-based algorithm for discovering clusters in large spatial databases with noise," in Kdd, 1996, vol. 96, no. 34, pp. 226-231.

[6] F. T. Liu, K. M. Ting, and Z.-H. Zhou, "Isolation forest," in 2008 Eighth IEEE International Conference on Data Mining, 2008, pp. 413-422: IEEE.

[7]   N. Chakrabarty, "A Gaussian Approach to the Detection of Anomalous Behavior in Server Computers," 2019.

[8]   B. Wang and Z. Z. Mao, "Outlier detection based on Gaussian process with application to industrial processes," (in English), Applied Soft Computing, vol. 76, pp. 505-516, 2019.

[9]   Z. Y. Algamal, "Diagnostic in Poisson Regression Models," (in English), Electronic Journal of Applied Statistical Analysis, vol. 5, no. 2, pp. 178-186, 2012.

Practical Handbook of Thermal Fluid Science, 2023, 36-59

HEAT TRANSFER

3.1. INTRODUCTION

Heat transfer is a subject that deals with the generation, use, conversion, and exchange of thermal energy between a system and its surroundings. At the molecular level, we can visualize thermal energy by the motion of particles and quantify it by the overall kinetic energy of all the particles. Heat transfer can then be explained as an exchange of the kinetic energy among particles *via* collisions or interactions. In a gas, the particles move faster, on average, under higher temperatures and transfer their kinetic energy to particles at lower temperatures through collisions. In a solid, the particles vibrate faster under higher temperatures and transfer their kinetic energy to particles at lower temperatures through molecular interactions.

We observe heat transfer in our everyday activities and many engineering developments. For example, the thermal energy released from burning natural gas in a burner on a kitchen stove heats a pan through convection, which is further conductively transferred to the food in the pan. In another example, on a hot summer day fans move air for body cooling via forced convective heat transfer: a higher air flow rate provides more cooling. The fan speed and power are well designed in factories to meet the cooling demand of our human body. In engine-driven vehicles, coolant circulates through internal combustion engines (ICE) for heat removal by forced convection. The coolant increases its temperature and delivers the heat to a radiator, where the heat is rejected to the ambient by both radiation and convection modes. Automakers design these systems to meet engine requirements, such as maximum temperature tolerance and heat removal rate while optimizing material costs and weight to meet vehicle-designed price points and fuel efficiency standards.

Inadequate heat removal causes material damage, device malfunction, and operational failures. For example, on February 1, 2003, the Space Shuttle Columbia disaster, a fatal incident in the United States space program that occurred when the Space Shuttle Columbia (OV-102) disintegrated as it reentered the atmosphere, was caused by the damage of a piece of foam insulation from the Space Shuttle external tank. During reentry, hot atmospheric gases penetrated the thermal shield and burned the internal wing structure [1]. In 2011, the disaster of the Fukushima Daiichi nuclear power plant in Japan was caused by an earthquake and tsunami, which shut down the electric power for cooling the reactor and removing the decay heat. The reactor

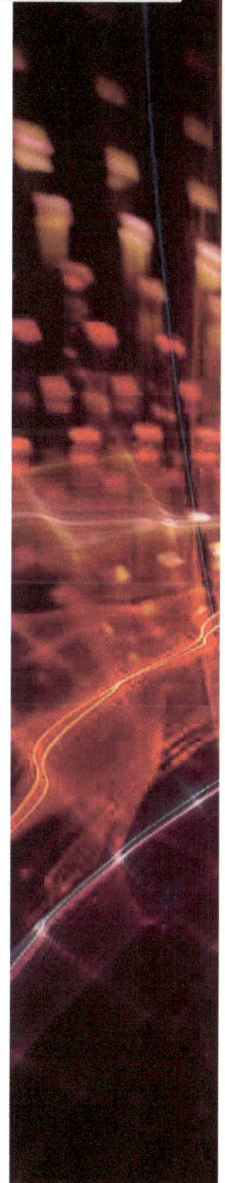

cores of units 1-3 overheated, which melted the nuclear fuel and caused leakage of contaminants and radioactive materials to the ambient. In an automobile ICE, leakage, low volume of coolant, or pump failure will reduce engine heat removal, causing high engine temperature and failure.

3.2. TEMPERATURE MEASUREMENT

Temperature is an important quantity widely used in our everyday activities. A body thermometer is used for fever diagnosis. The water of a lake or river freezes in a cold winter where the ambient temperature drops below 0 °C. Water boils at 100 °C, under one standard atmosphere (enough to raise a column of mercury 760 mm or pressure of 101,325 Pa). Hereinafter, we will refer to this pressure by the abbreviation 1 atm.

In thermodynamics, the temperature is a state property to evaluate the phase of a substance, internal energy, enthalpy, and heat transfer. In a Carnot cycle, thermal efficiency is directly expressed as a function of the high and low temperatures of the two associated reservoirs. The ideal gas law can be extrapolated to zero pressure, under which temperature drops to the absolute zero degrees (0 K or approximately -273.15 °C). In many industrial processes, the temperature is monitored to ensure the proper functions of devices and reactions. For example, state-of-the-art PEM fuel cells work under about 80 °C to produce power at high efficiency [2]. Too high or low temperature will reduce fuel cell performance and efficiency. In practice, thermocouples are equipped in a fuel cell system to monitor its thermal 'health' and ensure optimal operation [3].

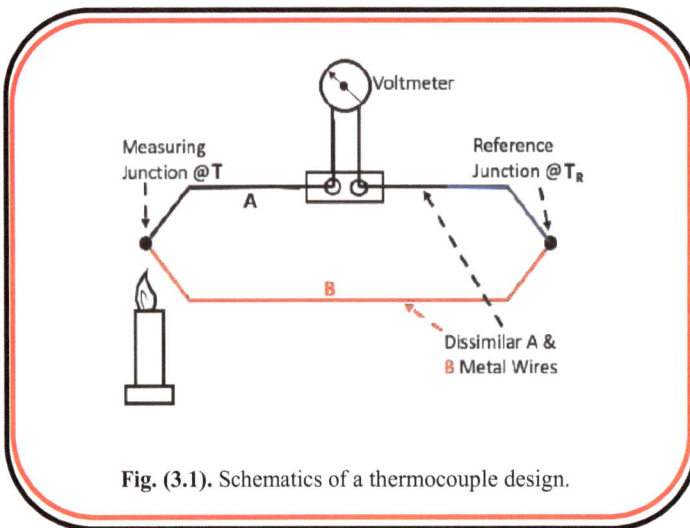

Fig. (3.1). Schematics of a thermocouple design.

Various types of thermometers are available commercially for temperature measurement. Mechanical thermometers, which use inherent characteristics of a material against a calibrated scale, are the first and primary thermometers. Liquid-in-tube thermometers are based on the principle of thermal expansion and measure the expanded volume of liquid in a narrow tube, which is then calibrated to temperature. Originally air was used, later alcohol, and then mercury, which expands less than alcohol, became the most common. With growing awareness of its toxicity, mercury thermometers began to be replaced with other types of liquid. Bimetallic thermometers are another popular type. They work by the differential expansion of two metals, commonly steel and copper, which causes the actuator to bend. This bending or twisting is then used to move a needle adjacent to a scale. Rugged and dependable, these thermometers are popular in industrial settings.

Electrically powered thermometers come in many types and are popular in the industrial, laboratory, and home settings. Resistance thermometers [4], also called resistance temperature detectors (RTD), are based on the temperature dependence of the material's electric resistance. A common RTD element structure consists of a fine wire wrapped around a ceramic or glass core. Platinum, nickel, and copper are popular wire materials. The voltage and current signals are directly measured to calculate the electric resistance, which is then calibrated to temperature. Because of the direct measurement of voltage and current signals, temperature output can be readily shown in the form of digital numbers, making it widely used in modern society. Their operation requires a power source and may be affected by its circuit resistance. Thermocouples are based on the thermoelectric Seebeck effect [5]: the temperature dependence of electric potential. A thermocouple measures the electric voltage between two junctions of dissimilar materials, which is converted to the temperature difference with a reference point T_R, as shown in Fig. (**3.1**). Because of direct voltage measurement, its operation is independent of the connecting wire length, interfacial resistance, and a power source, making it widely adopted in scientific and engineering studies. Table **3.1** lists popular types of thermometers.

Table 3.1. Equipment for the thermal-fluid experiment.

Types	Function	Image
Thermocouple	Measure a temperature-dependent voltage as a result of the thermoelectric effect, which is interpreted to temperature	

(Table 3.1) cont.....

Types	Function	Image
Liquid-in-tube Thermometer	Measure the liquid length, which changes with temperature due to thermal expansion.	
Infrared Thermometer	Measure temperature from a portion of the thermal radiation emitted by the object.	
Bimetallic Thermometer	Use a bimetallic strip to convert a temperature change into mechanical displacement due to different thermal expansions of the two metals	
Resistance Thermometer	Use wire windings or other thin-film serpentines that exhibit temperature dependence of electric resistance.	
Change-of-state Sensor	Measure the substance state that is temperature dependent. Commercial devices are often in the form of labels, pellets, crayons, or lacquers	
Silicon Diode	Measure the conductivity which increases linearly with temperature, usually for low cryogenic temperature.	

3.3. HEAT FLUX

Three modes of heat transfer are widely studied, namely heat conduction, heat convection, and heat radiation. They have distinct characteristics and may occur simultaneously in some occasions, *e.g.* cooking using a burner on a stove or in an oven and cooling of a nuclear reactor. Physically, heat conduction occurs through the microscopic exchange of the kinetic energy of constituent particles in objects. The flux is determined by the random motions of particles or molecules.

Heat convection is promoted by a bulk flow of a fluid (gas or liquid). Think about what is the difference between the random motions of particles and the bulk motion of fluids. The bulk flow can be induced by external forces such as fans and pumps or buoyancy from gravitational force. The former is known as forced heat convection, while the latter is natural convection. Fig. (**3.2**) schematically shows the difference between these two convective heat transfers.

Thermal radiation is transfer of energy by means of photons in electromagnetic waves, which can take place without physical contact. Both visible and invisible photons can transfer heat. Although a hot pan doesn't emit visible light, one can feel the heat from it with a hand held several centimeters above the surface. The heat emitted from reflected sunlight off a distant window is an example of visible spectrum (and invisible) thermal radiation.

The heat fluxes in these three transfer modes can be expressed as a function of temperature and fluid, material, or surface properties. Table **3.2** lists the heat flux expression for each mode of heat transfer.

Table 3.2 Heat flux for each mode of heat transfer.

Heat Transfer Mode	Heat Flux	Property
Conduction	$q_x = -k\dfrac{dT}{dx}$	k: thermal conductivity
Convection	$q = h(T_s - T_\infty)$	h: heat transfer coefficient
Radiation	$q = \epsilon\sigma(T_s^4 - T_\infty^4)$	ϵ: emissivity

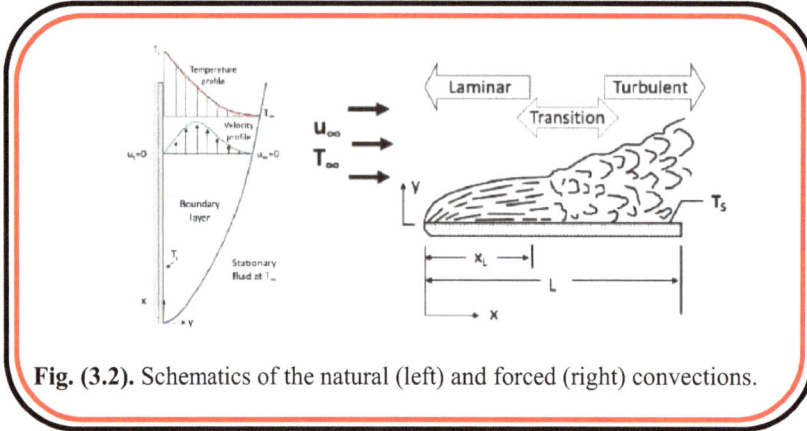

Fig. (3.2). Schematics of the natural (left) and forced (right) convections.

3.4. LUMPED HEAT CAPACITY ANALYSIS

In reality, heat transfer usually involves heat flow in all three physical dimensions, and thus temperature will vary in three dimensions (3-D). This 3-D variation makes it challenging to analyze heat transfer problems because mathematically, one needs to resolve a partial differential equation of heat transfer. In the case when only a small spatial variation is present, one can assume uniform temperature in the targeted object and then apply the lumped heat capacity analysis, also called "lumped capacitance model," to obtain useful information regarding temperature evolution or heat flow rate [6].

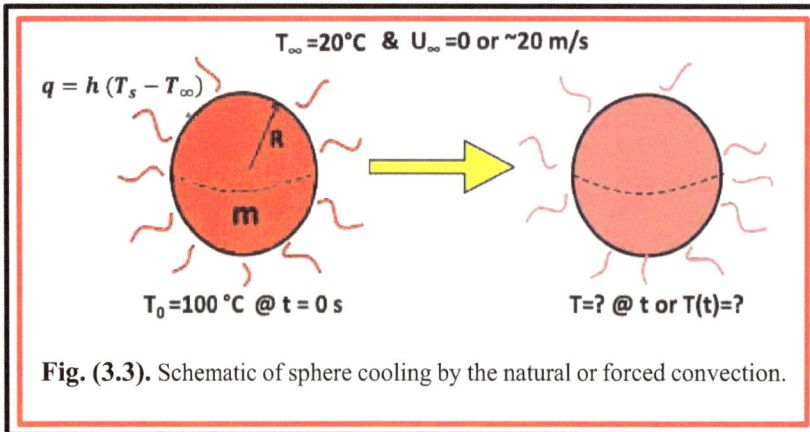

Fig. (3.3). Schematic of sphere cooling by the natural or forced convection.

To consider a simple cooling problem, a metallic ball, initially at 100 °C, is subject to air cooling by natural or forced convection at its surface, as shown in Fig. (3.3). An immediate simpliifcation can be made for the sphere of high thermal conductivity, such that the sphere cools without internal temperature gradients and spatial variation. Then, its temperature is assumed to be spatially uniform, though changing with time, and the lumped heat capacity assumption holds true. The energy balance between the rate of change of thermal energy and the surface heat flux times the area of the sphere gives.

$$-mC_v \frac{dT_v}{dt} = Q = qA_s \qquad\qquad\qquad\qquad 1$$

where m is the mass of the sphere, C_v the speciifc heat at constant volume, which is equal to that at constant pressure for a solid (C_p), T_v the bulk temperature of the sphere, t the time, A_s the surface area of the sphere, and Q the surface heat flow rate. The mass of the sphere is equal to its density times the volume. For the case of forced convection, the surface heat lfux is speciifed by Newton's "Law" of Cooling.

$$q = h(T_S - T_\infty) \qquad\qquad\qquad\qquad 2$$

where h is the heat transfer coeffcient, T_S is the surface temperature and T_∞ the bulk lfuid temperature. The heat transfer coefficient is actually an empirical parameter, not a universal constant (hence the quotes around "Law"). It is a function of the properties of the lfuid, the shape of the object, and the fluid lfow itself. Note that h is fundamentally different from the thermal conductivity k as used in Fourier's law. For natural convection, h varies as $(T_s - T_\infty)^n$, where n is an empirical number. For small temperature differences, Equation 2 is often applied as a good approximation, *i.e.*, n=1. Note that the orientation of the object relative to the gravity vector is also important in natural convection, but not for the symmetrical shape of a sphere. Typical values of h for air are 3-10 W/(m^2·K) for natural convection and 30-100 W/(m^2·K) for forced convection. Table **3.3** lists three methods to measure h.

Example 1: *Calculate q and Q for the brass sphere forced convection in Table 3.4 for h = 100 W/(m²·K) at t=50 s using Equation 2.*

Solution:

At t=50 s, T_s=85.9 °C and T_∞=22.7 °C,

$$q = h(T_s - T_\infty) = 6{,}320 \ W/m^2$$

For the sphere of R=2.54 cm,

the surface area: $A_s = 4\pi R^2 = 0.0081 \ m^2$

Then, $Q = qA_s = 51.3$ W.

Example 2: *Redo the calculation for the radiation heat flux using that in Table 3.2 and assuming the emissivity ϵ =1.*

Solution:

For ϵ =1 and the Stefan–Boltzmann constant $\sigma = 5.67 \times 10^{-8} \ W/m^2 \cdot K^4$

Note that T_s and T_∞ should be in the unit [K].

The radiation heat flux: $q = \epsilon\sigma(T_s^4 - T_\infty^4) = 508 \ W/m^2$

Then, $Q = qA_s = 4.12$ W.

Table 3.3 Methods of heat transfer coefficient measurement [7].

Method	Description	Equation
Direct Method	Based on Newton's "law" of cooling.	$h = \frac{q''}{T_s - T_\infty} k:$
Transient Method	Based on unsteady heat transfer and lumped capacity assumption.	$-hA_s(T_s - T_\infty) = mC_v \frac{dT_s}{dt}$
Wilson Method	For more than one convective heat transfer process, *e.g.* heat exchangers	$R_T = \frac{1}{h_i A_i} + \frac{\ln(\frac{d_e}{d_i})}{2k_w L_w} + \frac{1}{h_e A_e}$

Because of the lumped capacitance assumption, one can set $T_v = T_s$ in Equation 1, and thus with Equation 2, obtain:

$$-mC_v \frac{dT_s}{dt} = hA_s(T_s - T_\infty) \qquad \qquad \textbf{3}$$

Let us consider a constant T_∞ and assume that C_v is independent of temperature; one can solve the ifrst-order ordinary differential equation of Equation 3. To facilitate the application of mathematical methods, a dimensionless temperature θ is introduced, defined as:

$$\theta = \frac{T_s - T_\infty}{T_i - T_\infty} \qquad \qquad \textbf{4}$$

where T_i is the initial temperature of the sphere.

Example 3: Calculate θ for the brass sphere in Table 3.4 at t=50 and 100 s, respectively, using Equation 4 for forced convection.

Solution:

$T_i = 99.6\ °C.$

At t=50 s, $T_s = 85.9\ °C$ and $T_\infty = 22.7\ °C$, then θ = 0.822

At t=100 s, $T_s = 75.3\ °C$ and $T_\infty = 22.8\ °C$, then θ = 0.683.

Note that θ will therefore vary between 1 and 0 when cooling from T_i to T_∞. Equation 3 then becomes:

$$\frac{d\theta}{\theta} = d(\ln\theta) = -\frac{hA_s}{mC_v}dt \qquad\qquad 5$$

Since h is assumed independent of temperature or θ, Equation 5 integrates to:

$$\theta = C_1 e^{-\frac{hA_s t}{mC_v}} \qquad\qquad 6$$

where C_1 is the constant of integration. Using the initial condition θ=1 at t=0, C_1=1. Therefore, the mathematical solution to the cooling of the sphere is given by:

$$\theta = e^{-\frac{hA_s t}{mC_v}} \qquad\qquad 7$$

From the above solution, one sees that the temperature everywhere inside of the cooling sphere falls exponentially with time. The heat transfer coefficient h can then be determined from the rate of temperature decrease since the quantities A_s, m and C_v are known for a given sphere.

In addition, Equation 7 predicts that the time decay of the dimensionless temperature is exponential. Exponential decay of temperature, voltage, *etc.*, is common in many engineering problems. The usual form, in the context of low Biot number problems, is:

$$\theta = e^{-\frac{t}{\tau}} \text{ or } \ln\theta = -\frac{t}{\tau} \qquad\qquad 8$$

where τ is the time constant ($\tau = \frac{mC_v}{hA_s}$) in this cooling problem. In physics and engineering, the time constant is the parameter characterizing the response of a system to a step input. Good experimental practice is to allow the system to decay for at least one time constant, such that:

$$\theta = e^{-1} \cong \frac{1}{3} \qquad \qquad \longrightarrow \qquad 9$$

Example 4: For the brass sphere in Table 3.4, calculate the time constant for h equal to 1 and 100 W/m^2 K, respectively.

Solution:

For the Brass sphere,

$C_v = 385$ *J/kg K*, $A_s = 0.00811$ *m^2*, *m=0.585 kg,*

h=1 W/m^2 K $\rightarrow \tau = mC_v/hA_s = 27,800$ *s~ 8 hours*

h=100 W/m^2 K $\rightarrow \tau = 278$ *s ~ 5 min*

Since $\frac{t}{\tau}$ is dimensionless, one can cast it in the form of products of dimensionless quantities. We define a characteristic length for an object as the ratio of the volume to the surface area:

$$L_c = \frac{V}{A_s} \qquad \qquad \longrightarrow \qquad 10$$

where V is the volume of the object.

Example 5: Calculate L_c for the Aluminum sphere in Table 3.5. Derive L_c for a spherical ball and a cube.

Solution:

For the aluminum sphere, R=2.54 cm,

$V= 6.84 \times 10^{-5} m^3$ *and* $A_s=0.00811 m^2$ →$L_c=0.847 cm$

For a general spherical ball of R,

$$V=\frac{4}{3}\pi R^3 \text{ and } A_s=4\pi R^2 → L_c=R/3$$

For a cube of a side length L,

$$V=L^3 \text{ and } A_s=6L^2 → L_c=L/6.$$

Given that $m = \rho V$, one obtains

$$\frac{t}{\tau} = \frac{hA_s t}{\rho V C_v} = \frac{ht}{\rho L_c C_v} = \frac{hL_c}{k}\frac{k}{\rho C_v}\frac{t}{L_c^2} = \frac{hL_c}{k}\frac{\alpha t}{L_c^2} \qquad \textbf{11}$$

where α is the thermal diffusivity of the material of the object $k/\rho C_v$.

The dimensionless quantity hL_c/k is known as the Biot number and is a measure of the relative importance of convective heat transfer at the surface to conduction in the solid object:

$$Bi = \frac{hL_c}{k} \qquad \textbf{12}$$

For the lumped capacitance approximation to be valid, the thermal conduction in the object must be very large or the heat transfer to the fluid small expressed as $Bi \ll 1$. For a crude evaluation, $Bi < 0.1$ is sufficient for the lumped capacitance assumption. For Bi of about 1 or much larger, the spatial temperature variation inside the sphere will be considerable and the lumped

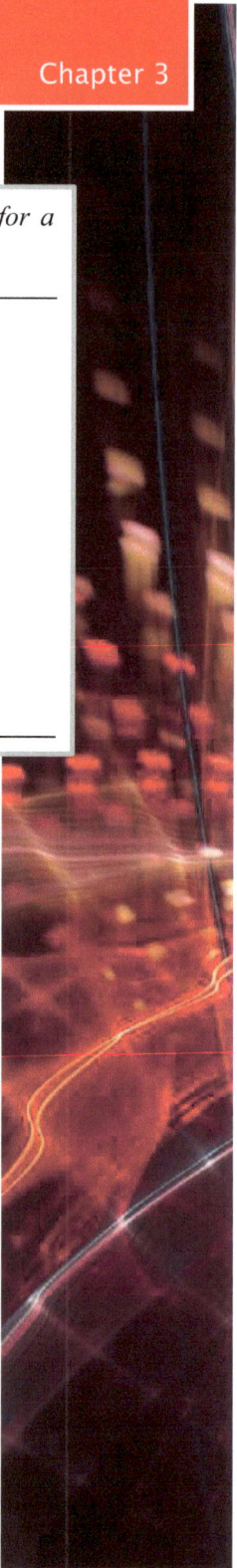

capacitance assumption fails. In this case, the changing rate of the thermal energy in the sphere in Equation 1 will be in the form of integral over the entire sphere's volume.

Example 6: Calculate the Biot number for the two spheres in Table 3.4 for h equal to 1 and 100 W/(m²·K), respectively.

Solution:

Formula/ Material	Brass		Aluminum	
	h=1 W/(m²·K)	h=100 W/(m²·K)	h=1 W/(m²·K)	h=100 W/(m²·K)
$Bi = hL_c/k$	7.63×10^{-5}	7.63×10^{-3}	3.57×10^{-5}	3.57×10^{-3}

3.5. EXPERIMENT

Unlike the thermal conductivity k that is an inherent material property, the heat transfer coefficient h is related to fluid flow, determined by a few parameters including the flow velocity, fluid thermal and transport properties (*e.g.* conductivity and viscosity), and object geometry. Table **3.3** shows three popular methods for h measurement. In this section, an example of an experiment is introduced to determine the heat transfer coefficient h based on the above 'lumped capacitance' analysis, *i.e.* the second method in Table **3.3**. First, a metal sphere is heated to a uniform temperature of... . °C in boiling water under 1 atm. Note that boiling water temperature is a function of the ambient pressure. At time t = 0, the sphere at 100 °C is exposed to ambient air cooling. A thermocouple is installed in the sphere to monitor the sphere temperature *versus* time. As shown in Equation 7, the heat transfer coefficient h can be determined from the temperature evolution data. Both natural and forced convections will be tested for air cooling. A sample of data is shown in Table **3.4** for cooling of the Brass and Aluminum spheres, obtained using the below experimental apparatus and procedure.

Table 3.4. Sample of the experimental data for sphere cooling.

Time	Brass Forced Conv.		Brass Natural Conv.*		Aluminum Forced Conv.*		Aluminum Natural Conv.*	
(sec)	T_s (°C)	T_∞ (°C)	T_s (°C)	T_∞ (°C)	T_s (°C)	T_∞ (°C)	T_s (°C)	T_∞ (°C)
0	99.641	22.666	94.778	24.002	98	23	98	23
10	97.318	22.658	94.309	24.010	96	23	96	23
20	93.545	22.665	93.842	23.961	93	23	93	23
30	90.916	22.589	93.386	23.952	92	23	92	23
40	88.328	22.626	92.930	23.940	91	23	91	23
50	85.936	22.669	92.471	23.913	91	23	91	23
60	83.637	22.740	92.046	23.937	88	23	90	23
70	81.437	22.846	91.587	23.880	84	24	90	23
80	79.323	22.917	91.153	23.857	81	24	89	23
90	77.254	22.881	90.738	23.782	77	24	89	23
100	75.268	22.828	90.325	23.694	73	24	88	23
110	73.380	22.828	89.923	23.602	71	24	87	23
120	71.491	22.836	89.553	23.635	68	24	87	23
130	69.646	22.876	89.233	23.722	65	24	86	23
140	67.947	22.788	88.914	23.738	63	24	86	23
150	66.300	22.791	88.588	23.703	60	24	86	23
160	64.718	22.802	88.262	23.718	58	24	85	23
170	63.217	22.820	87.949	23.732	56	24	85	23
180	61.778	22.961	87.646	23.687	54	24	84	23
190	60.384	23.116	87.351	23.553	52	24	84	23
200	59.043	23.199	87.050	23.475	50	24	84	23
210	57.738	23.263	86.756	23.453	49	24	83	23
220	56.481	23.320	86.471	23.355	47	24	83	23
230	55.287	23.343	86.184	23.282	46	24	82	23
240	54.126	23.398	85.892	23.241	44	24	82	23
250	53.018	23.509	85.598	23.171	43	24	82	23
260	51.959	23.551	85.322	23.136	42	24	81	23
270	50.920	23.579	85.052	23.102	41	24	81	23
280	49.918	23.537	84.775	23.073	40	24	81	23
290	48.929	23.573	84.510	23.067	39	24	80	22
300	47.995	23.587	84.238	23.070	38	24	80	22
310	47.081	23.568	83.976	23.078			80	22
320	46.212	23.585	83.715	23.074			79	22
330	45.374	23.639	83.462	23.047			79	22
340	44.562	23.633	83.188	23.022			78	22
350	43.812	23.650	82.913	22.994			78	22
360	43.077	23.634	82.657	22.944			78	22
370	42.378	23.543	82.410	22.935			77	22
380	41.689	23.653	82.155	22.935			77	22
390	41.012	23.703	81.900	22.883			77	22
400	40.371	23.761	81.638	22.778			77	22
410	39.742	23.650	81.364	22.736			76	22
420	39.109	23.692	81.101	22.760			76	22
430	38.532	23.763	80.846	22.784			76	22

*Data recording starts with a temperature lower than the boiling point.

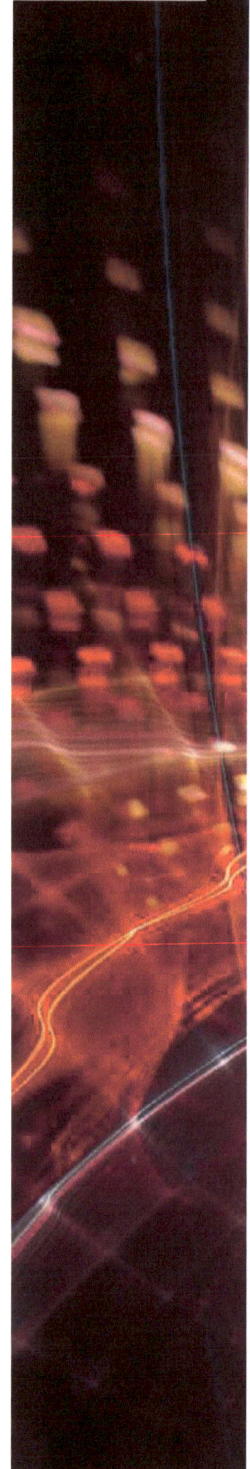

3.5.1. Apparatus

The apparatus consists of brass and aluminum spheres with type K thermocouples embedded in their centers. Another thermocouple monitors the ambient temperature. The thermocouples are connected to a National Instruments' data system. The physical properties and dimensions of the spheres are given in Table **3.5**.

Table 3.5. Configuration of the spheres in experiment.

Property	Brass	Aluminum
Thermal Conductivity	111 W/mK	237 W/mK
Specific Heat Capacity C_v	385 J/kgK	903 J/kgK
Density	8,520 kg/m³	2,702 kg/m³
Thermal Diffusivity	3.384×10^5 m²/s	9.71×10^5 m²/s
Sphere Radius	2.54 cm (1 inch)	2.54 cm (1 inch)

How long should the temperature-time data be taken? For the theoretical limit of equilibrium of the sphere temperature with the ambient environment, the waiting time will be infinite, as shown in Equation 8. In the experiment, one could use an arbitrary rule, like 50% of the initial temperature or something equivalent, but the theory of the time constant should enable us to make a more rational choice. In general, a duration of about one time constant is desirable in the experiment, as shown in Equation 9. One can then evaluate the pre-estimated final temperature in the experiment. This should be calculated prior to the measurement.

#Example 7: Calculate the temperature of the aluminum sphere after one time constant when cooling from 100 °C by 20 °C air.

Solution:

Formula	$T_s(\tau = 1) = e^{-1}(T_i - T_\infty) + T_\infty$
T_i [°C]	100
T_∞ [°C]	20
T_s [°C]	49

3.5.2. Procedure

The National Instruments' data program will be used for temperature-time data acquisition. It is recommended to conduct a test on data recording to ensure that the temperatures of the thermocouples can be retrieved and saved to a file in a computer. In experiment:

Step: 1 Boil Water

Add tap water to the pot, place the free thermocouple in the water so it does not touch the pot wall, start the data acquisition program, and then turn on the hot plate. Watch the time-temperature plot in the monitor until the water boils vigorously.

Example 8: What is the boiling temperature of water under 1 atm? Estimate the boiling temperature in Denver, Colorado, which is about 1.6 kilometers (1 mile) above sea level.

Solution:

From Steam Tables, the boiling temperature is 100 °C under 1 atm.

In Denver, the atmospheric pressure is lower than 1 atm, which can be estimated by the altitude.

For g=9.8 m/s², ρ=1.2 kg/m³, and h=1,600 m

$$P = \rho g h = 18{,}800 \ Pa$$

Denver's atmospheric pressure: P_{atm}=101,300-18,800 = 82,500 Pa

From Steam Tables, the boiling temperature is about 94 °C under 82,500 Pa.

Step: 2 Boil a Sphere

This run will permit the determination of the heat transfer coefficient between a sphere and boiling water. One can derive it following the same procedure of the sphere cooling in Equations 1-7. Note that the heat transfer coefficient of boiling may be very large, so one needs to evaluate the Bi number to make sure the analysis is valid, *i.e.* Bi <<1. In an experiment, one sphere initially at the ambient temperature is quickly plunged into the boiling water until reaching the boiling temperature. Note the sphere should not touch the pot wall or bottom during heating. The pot wall usually has a different temperature from the boiling water due to the thermal resistance between the liquid and solid wall.

Step: 3 Conduct Sphere Cooling

In this step, the hot sphere will be subject to the forced or natural convection in ambient air.

(a) Forced Convection

When the sphere reaches the boiling temperature, i) move the ambient thermocouple from the pot and place it in a controlled air flow, ii) set the air speed to ~20 m/s, iii) start the data recording, iv) quickly remove the sphere by the handle from the pot and place it in the air flow, and v) record the temperature data until the pre-estimated final temperature of the sphere is reached.

(b) Natural Convection

When the sphere reaches the boiling temperature, i) place the ambient thermocouple in the vicinity of the ring stand, ii) start the data recording, iii) quickly remove the sphere by the handle from the pot and place it on the ring stand away from any air currents, and iv) record data until the pre-estimated final temperature is reached.

3.5.3. Summary of Experimental Runs

Seven data runs can be obtained and summarized below:

Run #	Sphere	Condition
1	T/C	Boil
2	Br or Al	Heat in boiling water
3	Al or Br	Heat in boiling water
4	Br or Al	Cool air forced
5	Al or Br	Cool air forced
6	Br or Al	Cool air natural
7	Al or Br	Cool air natural

The temperature evolution of a sphere in the testing from boiling water heating to air cooling is schematically shown in Fig. (**3.4**), where the sphere is dropped to the boiling water at Point 1 and the air cooling starts at Point 2. A complete sample of data is available at the website of book materials.

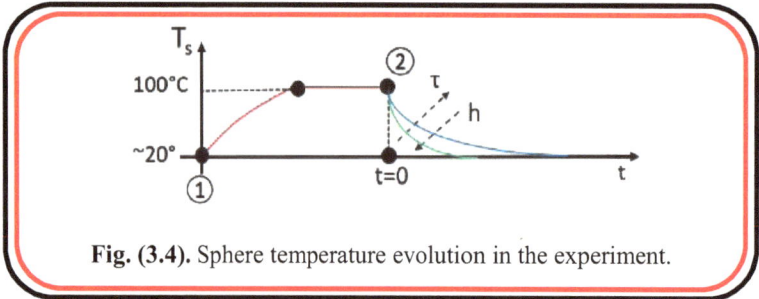

Fig. (3.4). Sphere temperature evolution in the experiment.

3.5.4. Data Analysis

a.) ln θ *versus* Time t

In the sphere cooling, one can calculate θ_i and $\ln \theta_i$ from the tabulated data of T_i at t_i. Because $\ln \theta$ and time t follow a linear relationship as shown in Equation 8, one can plot $\ln \theta$ *versus* time t, and find the line that best fits all the data. Note that in order to obtain the linear relationship, the initial time must be set to be 0 when the cooling is initiated, *i.e.*, the sphere is placed in an air flow or ambient air. Thus, proper data post-processing is required to shift the time to ensure this initial condition. Fig. (**3.5**) shows the relationship between $\ln \theta$ and time t and its relevance to the time constant. Fig. (**S3.4**) plots

the profiles using the experimental data of natural convection cooling.

In the sphere heating by the boiling water, similar analysis can be developed following the cooling method as long as the associated Da number is small.

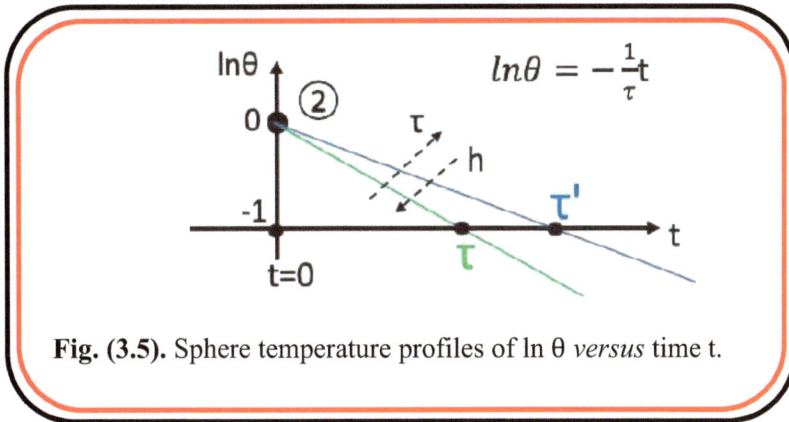

Fig. (3.5). Sphere temperature profiles of ln θ *versus* time t.

b.) Heat Transfer Coefficient *h*

To obtain the heat transfer coefficient *h* from the data, one can calculate the slope of the fitting line. The slope is a function of *h*, as shown in Equation 8, which then enables the determination of *h*. Note that the straight line is the best fitting for all the data in the cooling; thus the obtained *h* can be treated as the final result from the measurement.

 # Example 9: Calculate h from the slope of the two data at 50 and 100 s for the brass sphere under force convection in Table 3.4, using Equation 8.

Solution:

Set (x, y) = (t, ln θ): for the two data points, (50s, -0.196) and (100s, -0.381)

$Slope = \dfrac{y2 - y1}{x2 - x1}$	$h = -slope(\dfrac{mC_v}{A_s})$
-0.00371 [1/s]	103 $[W/m^2 K]$

In addition, mathematically, any data point (T_i, t_i) in the cooling curve can be used to calculate the heat transfer coefficient h_i, as shown in Equation 7. As a result, a set of h_i will be obtained from all the data points. The average of

$\{h_i\}$ can be taken to compare with the one obtained from the slope. Note that one could calculate the standard deviation of $\{h_i\}$ and plot the frequency distribution. However, it is unlikely the data set of $\{h_i\}$ will follow the Gaussian distribution.

Example 10: Calculate h from the data at 100 s for the brass sphere under force convection in Table 3.4 using Equation 7.

Solution:

Formula	$h = \dfrac{mC_v \ln\theta}{-tA_s}$
C_v [J/kg K]	385
A_s [m^2]	0.00811
t [sec]	100
ln θ	-0.386
m [kg]	0.58483
h [W/m^{2}K]	**107**

3.6. 3-D NUMERICAL ANALYSIS

To enhance the fundamental understanding of the above cooling problem and analysis, heat transfer simulation software can be employed to predict the temperature 3-D distribution and evolution under the experimental conditions. The 3-D temperature information is obtained by numerically solving the heat transfer governing equation, which is mathematically a partial differential equation (PDE):

$$\frac{\partial^2 T}{\partial x^2} + \frac{\partial^2 T}{\partial y^2} + \frac{\partial^2 T}{\partial z^2} = \frac{\rho C_p}{k}\frac{\partial T}{\partial t} \qquad \qquad 13$$

The heat transfer coefficient h is present in the boundary condition at the sphere surface:

$$-k\frac{\partial T}{\partial r} = h(T - T_\infty) \qquad \qquad 14$$

Note that the sphere cooling in the above experiment is essentially an unsteady-state 1-D problem, *i.e.* temperature only varies with time and radius r only. The spherical coordinate is preferred in the numerical problem setup. The heat transfer equations in the three typical coordinates are documented in Appendix III. In the 3-D simulation result; no temperature variation should develop in the other two directions. SolidWorks is capable of 3-D heat transfer simulation. A brief introduction of how to set up a case is given in the website of additional book materials. In 3-D simulation, one can change the sphere size, thermal conductivity, and heat transfer coefficient to study impacts of different values of Da, *e.g.* 0.1, 1, and 10. Additionally, one can use the exact experimental condition, including the calculated h, to predict the temperature evolution and compare it with the experimental data.

For further readings regarding heat transfer and experiment, we refer the interested readers to the references in [8-13].

3.7. QUESTIONS

1. In the sample of experimental data, estimate the uncertainties in the measurements of time and temperature.

2. Plot T_s *versus* t and ln θ *versus* t for each run. Several runs on each plot, with legend or multiple small plots on one page, are OK. Clearly show the values on the ln θ versus t plots and find a way to use the curve of ln θ *versus* t to calculate h.

3. Theoretically, you can use any point (T_s, t) to calculate h. From the development of Equation 7 and the experimental design, what do you think may contribute to the errors (not related to the exp. equipment) in calculating h? How to minimize the errors?

4. Compute the Biot number for each run. Are the Biot numbers small enough to justify the lumped capacity system assumption?

5. Do you expect the heat transfer coefficients h to be different for the brass and aluminum spheres under the same conditions? If yes, why? If no, why? Compare your results for the brass and aluminum spheres and comment.

6. Compare your values of h to those in heat-transfer references.

7. From the basic measurements, if using Equation 7 and any point (T_s, t) to calculate h, estimate δh using δT and δt. Please comment on how to minimize δh.

8. Use Solidworks to reconstruct a sphere and run the simulation and plot the temperature profile change with time (T(r,t)). Use one data set of cooling to compare with the analytical result (Equation 8) and simulation result (using the center temperature). For simulation, reduce the thermal conductivity by 100, please replot T(r,t) and comment on the result.

9. The TES tank at the central plan of UC IRVINE is a large cylinder 105 feet high and 88 feet in diameter to store the excess chilled water. Estimate the heat loss from the tank for winds of 0 and 20 m s^{-1}. Take the surface of the tank to be 40 °F and the air temperature to be 120 °F. For a 12 hour period, compare energy loss to "that" stored in the tank. Would more insulation be worth it? Use the heat transfer coefficients from the experiments.

10. Use ~30 data points around the time constant in your measurement of forced convection for the brass sphere cooling to calculate h (~30). Plot the frequency distribution of h (please choose a proper bin size). Make a comment on the Histogram.

11. If the cooling starts at $t=10$ s, instead of 0 s, please rederive the solution to Equation 1.

12. In Table **3.4**, comment on whether the precisions or data recording rate (*i.e.* one data per 10 s) are reasonable.

13. In Table **3.4**, the T_∞ data are shown to be changing with time. Comment on its change with time!

14. In Table **3.4**, some T_i data start with a temperature well below the boiling point. Comment on the impact on the h evaluation.

15. In Table **3.4**, for the first set of data please calculate h using the sub-set of data, starting with t=50 s, *i.e.* the initial temperature is the temperature at t=50 s, at which the cooling starts. Comment on the result!

16. In Table **3.4**, for the first set of data, using the maximum and minimum T_∞ data, respectively, as T_∞ for the h calculation. Comment on the results!

17. What assumptions would you make to analytically solve for the effectiveness (measured in terms of its insulating properties) of a double pane *versus* a single pane window? Give relevant equations.

REFERENCES

[1] "6.1 A History of Foam Anomalies," 2003, Available: https://web.archive.org/web/20110516132723/ http://caib.nasa.gov/news/report/pdf/vol1/chapters/chapter6.pdf.

[2] Y. Wang and M. Gundevia, "Measurement of thermal conductivity and heat pipe effect in hydrophilic and hydrophobic carbon papers". International Journal of Heat and Mass Transfer, vol. 60, pp. 134-142, 2013.

[3] Y. Wang and K. S. Chen, "PEM fuel cells: thermal and water management fundamentals". Momentum Press, 2013.

[4] C. H. Meyers, "Coiled filament resistance thermometers", NBS Journal of Research, vol. 9, 1932.

[5] T. Geballe and G. Hull, "Seebeck effect in silicon", Physical Review, vol. 98, no. 4, p. 940, 1955.

[6] F. P. Incropera, A. S. Lavine, T. L. Bergman, and D. P. DeWitt, "Fundamentals of heat and mass transfer". Wiley, 2007.

[7] T. A. Moreira, A. R. A. Colmanetti, and C. B. Tibiriçá, "Heat transfer coefficient: a review of measurement techniques", Journal of the Brazilian Society of Mechanical Sciences and Engineering, vol. 41, no. 6, p. 264, 2019.

[8] R. A. Granger, (Ed.). "Experiments in heat transfer and thermodynamics". Cambridge University Press, 1994.

[9]   A. F. Mills, "Basic heat and mass transfer". Prentice hall, 1999.

[10] W. M. Rohsenow, J. P. Hartnett and Y. I. Cho, Handbook of heat transfer (Vol. 3). New York: McGraw-Hill, 1998.

[11] S. Kakaç, R. K. Shah and W. Aung, "Handbook of single-phase convective heat transfer", 1987.

[12] A. Bejan and A.D. Kraus, (Eds.). "Heat transfer handbook" (Vol. 1). John Wiley & Sons, 2003.

[13] R. Wang, R. Wang, & Y. Wang, "Ex-situ Measurement of Thermal Conductivity and Swelling of Nanostructured Fibrous Electrodes in Electrochemical Energy Devices". Thermal Science and Engineering Progress, 100805, 2020.

Practical Handbook of Thermal Fluid Science, 2023, 60-80

POWER PLANT

4.1. INTRODUCTION

Power plants or generating stations are facilities for generating electric power, which has become essential to our present lifestyle. Electric power supports many of our daily activities, including lighting, air circulation, HVAC, TV watching, cell phones, computers, elevators, *etc*. The generated power is delivered using complex systems of wires, towers, underground stations, transformers, control stations, *etc*., to consumer sites such as homes, apartments, stores, industry, and schools. These interconnected systems are known as grids.

For the past 150 years, most power plants have used fossil fuels, such as coal, petroleum, and natural gas, to generate electricity. Heat engines convert thermal energy from fuel combustion to rotating kinetic energy to turn generators and produce electric power. Owing to their high efficiency, continuous smooth motion, durability, low maintenance cost, and the possibility of production in enormous size, steam turbines remain the most popular heat engines in traditional central-station large power plants.

In operation, boilers burn fossil fuels and produce high-temperature, high-pressure steam. The steam passes through a steam turbine for energy conversion and condenses at the condenser side. The liquid condensate is then pumped back to the boilers for steam production. Various designs have been developed to improve the steam turbine and overall conversion of thermal energy to electric power generation, including the combined-cycle configuration [1, 2].

In recent years, clean and renewable power generation methods, including nuclear, solar photovoltaic, solar thermal, geothermal, biomass, waste-to-energy, and wind, have received growing attention. Steam turbines are a major part of nuclear, geothermal, biomass, waste-to-energy, and solar thermal power plants. These clean energy sources produce high-pressure steam, driving steam turbines for energy conversion. Some renewable and clean electric power plants, such as solar panels, wind turbines, hydropower, and fuel cells, do not rely on heat engines for energy conversion.

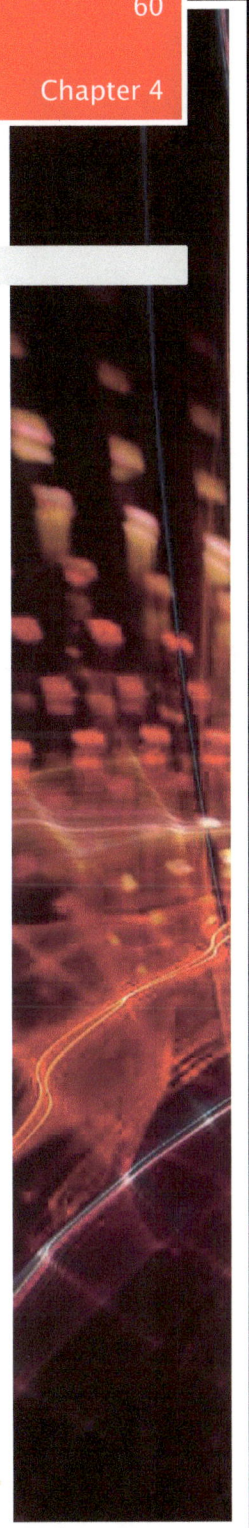

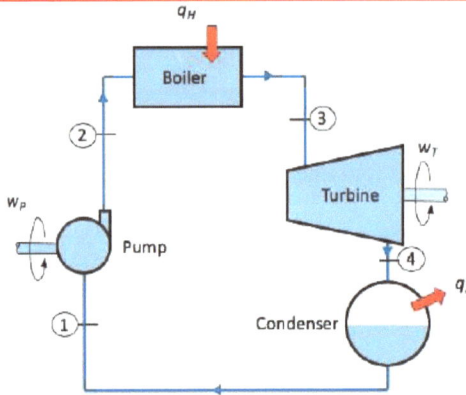

Fig. (4.1). Steam turbine and ideal Rankine cycle.

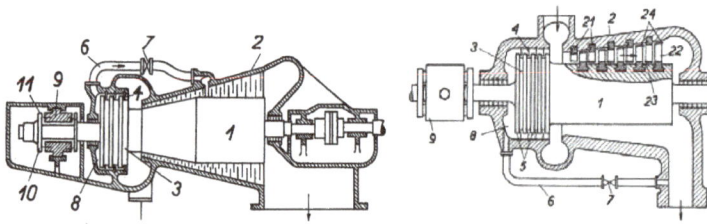

Fig. (4.2). Steam turbine design in a patent: 1 is the rotor and 2 is the casing of the turbine. 3 is a dummy piston for balancing the axial thrust acting on the blading of the shaft and 4 are the packing on the dummy piston [3]. Steam flows through the sealing gap between the packing members 4 and collars 5 on the shaft and pipe 6 to a lower stage of the turbine or condenser.

4.2. STEAM TURBINE POWER GENERATION

A steam turbine is a heat engine that extracts thermal energy from steam and transforms it into the mechanical work of a rotating shaft, which (in electric power generation) drives a generator to produce electricity. Steam turbine operation can be described by the Rankine Cycle, as shown in Fig. (**4.1**), with 4 processes and water/steam as its working fluid:

- Process 1-2: Isentropic compression in a pump. Water condensate is pumped from a state of low-pressure liquid to a high-pressure state.
- Process 2-3: Constant pressure heat addition in a boiler. High-pressure water enters a boiler, where heat is added by fuel combustion, nuclear reaction, or other sources, to vaporize the water under constant pressure. In water-tube boilers, the vapor in the drum is saturated as long as there is liquid water. The addition of heat to the steam beyond the drum will

increase its temperature, which is called superheating. Note that nuclear power plants do not generally superheat steam.

- Process 3-4: Isentropic expansion in a turbine. Superheated steam goes through multi-stage expansion in a steam turbine, which produces the mechanical work of shaft rotation. The steam pressure and temperature drop during this process.
- Process 4-1: Heat rejection under constant pressure in a condenser. Steam enters a condenser, where it condenses under constant pressure to become condensate, back to the initial state of the cycle.

Table **4.1** lists the Rankine cycle processes and major equations. Fig. (**4.2**) shows a steam turbine design proposed in a patent. The thermodynamic processes of the ideal Rankine cycle are shown by the T-s diagram in Fig. (**4.3**) (left). In power generation, the rotating shaft connects with an electric generator in the process 3-4.

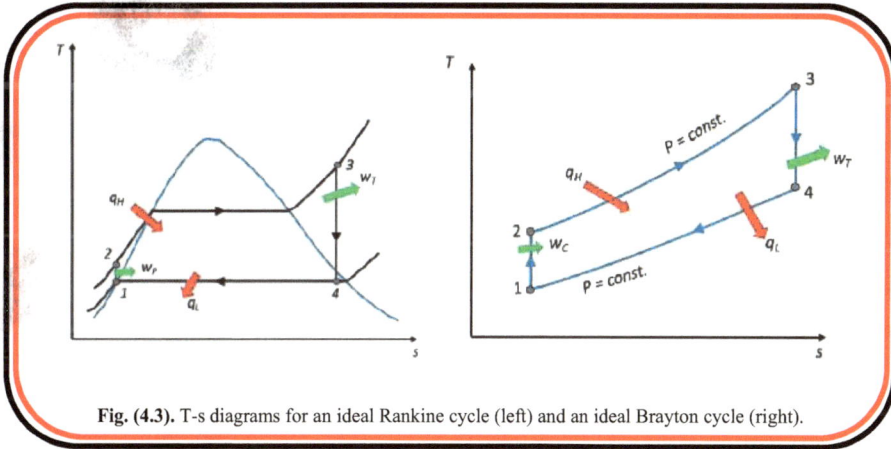

Fig. (**4.3**). T-s diagrams for an ideal Rankine cycle (left) and an ideal Brayton cycle (right).

Table 4.1. The Rankine cycle processes.

Component	Energy Eq.	Entropy Eq.	Process
Pump	$0 = h_1 + w_P - h_2$	$0 = s_1 - s_2 + (0/T) + 0$	$q = 0, s_1 = s_2$
Boiler	$0 = h_2 + h_3 - q_H$	$0 = s_2 - s_3 + \int dq/T + 0$	$P_3 = P_2 = C$

(Table 4.1) cont.....

Component	Energy Eq.	Entropy Eq.	Process
Turbine	$0 = h_3 - h_4 - w_T$	$0 = s_3 - s_4 + (0/T) + 0$	$q = 0, s_3 = s_4$
Condenser	$0 = h_4 - h_1 - q_L$	$0 = s_4 - s_1 - \int \frac{dq}{T} + 0$	$P_4 = P_1 = C$

Table 4.2. The Brayton cycle processes.

Component	Energy Eq.	Entropy Eq.	Process
Compressor	$0 = h_1 + w_C - h_2$	$0 = s_1 - s_2 + (0/T) + 0$	$q = 0, s_1 = s_2$
Combustor	$0 = h_2 + h_3 - q_H$	$0 = s_2 - s_3 + \int dq/T + 0$	$P_3 = P_2 = C$
Turbine	$0 = h_3 - h_4 - w_T$	$0 = s_3 - s_4 + (0/T) + 0$	$q = 0, s_3 = s_4$
Heat exchanger	$0 = h_4 - h_1 - q_L$	$0 = s_4 - s_1 - \int \frac{dq}{T} + 0$	$P_4 = P_1 = C$

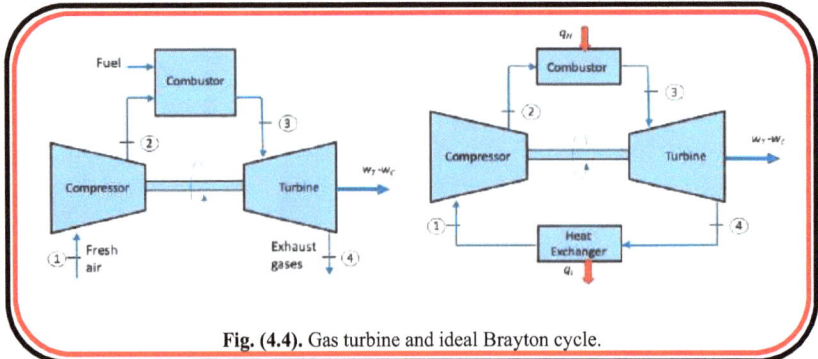

Fig. (4.4). Gas turbine and ideal Brayton cycle.

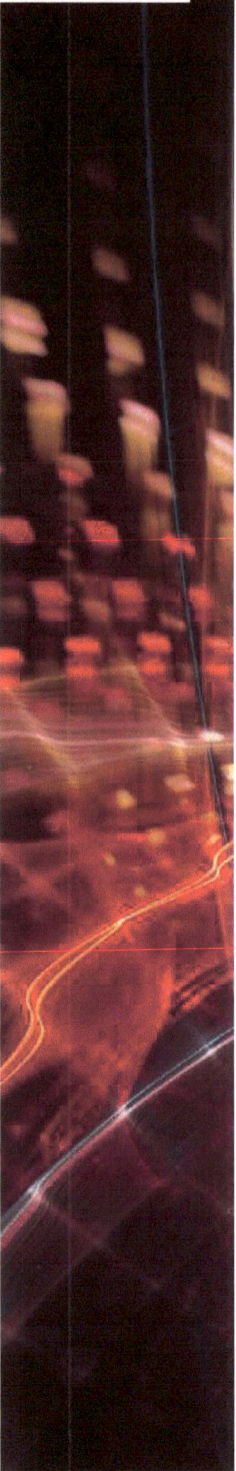

4.3. GAS TURBINE POWER GENERATION

A gas turbine (combustion turbine) is a heat engine that harnesses the thermal energy in fuel combustion to generate mechanical energy of rotary motion. Gas turbines can be connected for a mechanical drive (*e.g.* compressors) or electric power generation. Gas turbine operation is based on the Brayton Cycle with air as the working fluid, as shown in Fig. (**4.4**). The ideal cycle consists of four major processes:

- Process 1-2: Isentropic compression in a compressor. Ambient air is fed into a compressor, where it is pressurized with a slight increase in temperature.
- Process 2-3: Constant-pressure heat addition. Heat is added by spraying fuel into the compressed air for combustion, which generates a high-temperature flow.
- Process 3-4: Isentropic expansion in a turbine. The high-temperature flow then enters a turbine, where it expands to the ambient pressure and produces a shaft work. A portion of the shaft work is used to drive the air compression in Process 1-2.
- Process 4-1: Constant-pressure heat rejection. The turbine exhaust gas rejects heat to the ambient and returns back to its initial state condition.

Table **4.2** lists the Brayton cycle processes and major thermodynamics equations. The thermodynamic processes of an ideal Brayton cycle are shown in the T-s diagram of Fig. (**4.3**) (right). The shaft work in Process 3-4 drives an electrical generator in gas turbine power generation.

#Example 1: A 10 MW steam turbine power plant has an efficiency of 31.5% with its condenser's cooling water: $T_{in} = 60°F$ and $T_{out} = 110°F$, please

a) *Estimate the rate of heat rejection in the condenser (MW).*

b) *Estimate the mass flow rate of the condenser cooling water $\left(\frac{lbm}{hr}\right)$.*

How many pools (in 10×25 m^2) can the rejected heat support in winter of 10°C ambient air for constant 27°C pool water, assuming its heat loss is only via the pool surface at a heat transfer coefficient $h = 100 \frac{W}{m^2°C}$?

Solution:

(a).

Formula	$Q_H - Q_L = W_T$
η_T	0.315
W_T [MW]	10
Q_L [MW]	**21.75**

(b).

Formula	$Q = \dot{m}Cp\Delta T$
T_{in} [K]	288.706
T_{out} [K]	316.483
Q_L [W]	21750000
C_p of water $\left[\frac{J}{kg\ K}\right]$	4187
$\dot{m}$	$187 \left[\frac{kg}{s}\right]$ **or** $1480000 \left[\frac{lb}{h}\right]$

(c).

Formula	$Q_L = nAh(T_W - T_\infty) \left[\frac{W}{m^2}\right]$
T_∞ [°C]	10
T_W [°C]	27
Q_L [W]	21750000
$h \left[\frac{W}{m^2\ °C}\right]$	100
A [m^2]	250
n	**51**

4.4. CHILLER AND THERMAL ENERGY STORAGE TANK

Chillers, widely used in commercial and industrial facilities, are a thermodynamic device that produces chilled liquid from a refrigeration process. Most of them are based on the vapor-compression refrigeration—a reversed Rankine cycle, which requires power input to reverse the heat transfer direction. A design is shown in Fig. (4.5), which connects its condenser with a cooling tower for heat rejection. The working fluid is refrigerant, which vaporizes in the evaporator and absorbs heat from the liquid to be chilled. The coefficient of performance (COP) is one major measure used to evaluate chiller efficiency. Another measure using mixed units is kilowatts per ton (kW/ton), which is a handy method as electricity usage can be figured directly.

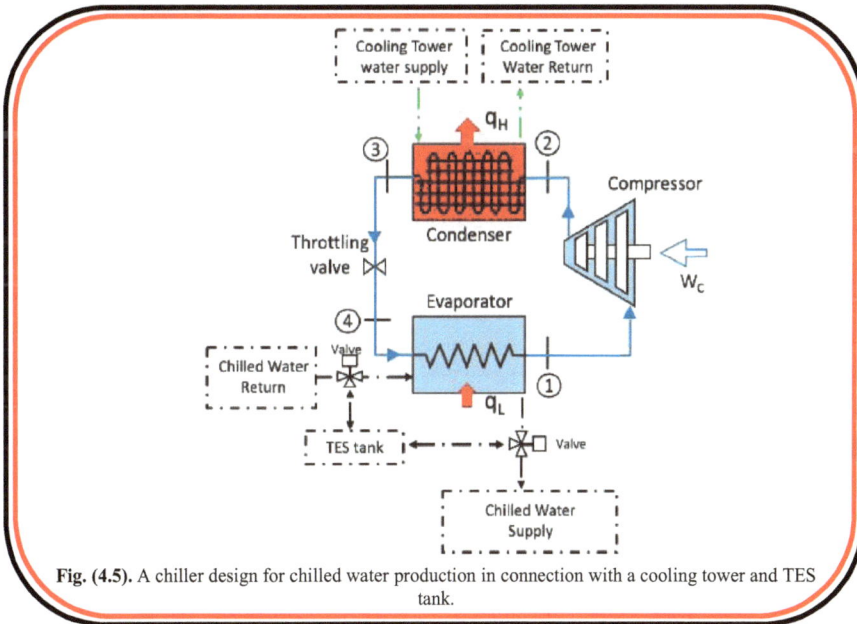

Fig. (4.5). A chiller design for chilled water production in connection with a cooling tower and TES tank.

Thermal energy storage (TES), often used in renewable energy applications, stores thermal energy in a medium in an insulated space so that it can be used later for heating, cooling or power generation. Fig. (4.6) shows a design example of a thermal energy system in the residential application using the heat extracted from compressed gas in a refrigerating cycle plus additional heat from any suitable heat dispensing means such as from a fuel-burning furnace coupled with ice storage produced by the refrigerating cycle. The design consists of a refrigerating cycle comprising a compressor 10 to compress the gas to its critical pressure. Heat in the compressed refrigerant is transferred to the water flowing through pipe 15, tank 13, and pipe 16, which is thereby heated to about 130 to 140°F. Assuming that the operation takes place in cold seasons, the water will be directed into coil 17 by opening valves 18 and 19, while coil 20 is kept inactive by keeping relevant valves closed. The coil 17 is preferably located in an air duct through which air coming from outside or air that is recycled from space or rooms being heated passes in the direction of the arrows. To supply more heat supply in cold weather, there is an auxiliary heat supplying device consisting of a boiler 23 and a steam coil 24, which is located in a duct leading to the space to be heated.

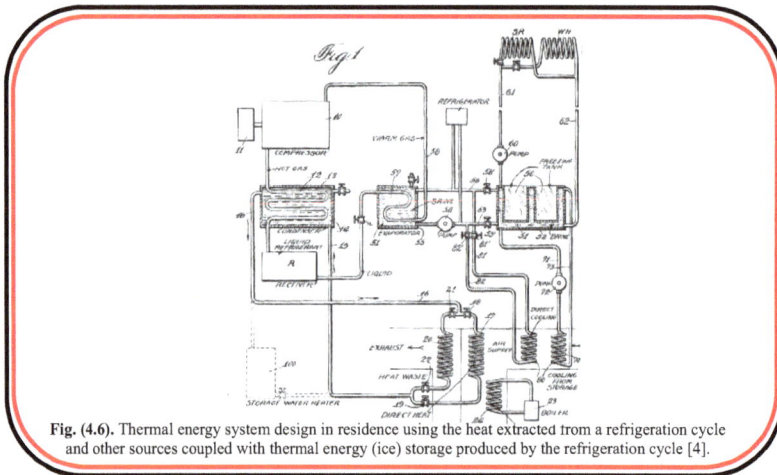

Fig. (4.6). Thermal energy system design in residence using the heat extracted from a refrigeration cycle and other sources coupled with thermal energy (ice) storage produced by the refrigeration cycle [4].

4.5. BALANCE OF PLANT

The Balance of Plant (BOP) is the supporting components and auxiliary systems in a power plant, other than the generating unit itself. BOP provides essential support to keep the plant operating. In a power plant, BOP consists of both electrical and mechanical devices. Electrical BOP devices include main transformers, auxiliary transformers, circuit breakers, switchgear, surge arresters, electrical buss bars, etc. Mechanical components of BOP include pumps, water treatment, controls, instruments, fuel conditioning systems, pressure regulation systems, NO_x reduction units, fire protection systems, air compressors, *etc.* During plant design, mechanical and electrical BOP components are optimized to increase the efficiency of power plant operations and to reduce operational downtime.

4.6. MAJOR COMPONENTS IN A POWER PLANT

In a traditional power plant, electricity is produced, along with waste heat rejection, heat exchange, steam production/condensation, and water circulation. A comprehensive power plant can be developed to fully use the existing energy such as electricity, waste heat, and steam for multiple functions that benefit our human activities. As an example, the Central Plant (CP) at the UC Irvine is introduced in this section, which integrates multiple functions of gas turbine power generation, steam turbine power generation, hot water production, chilled water production, and thermal energy storage. Several major subsystems and components are listed below:

1. High Temperature Water (HTW) supply and return pipelines
2. Chilled Water (CW) supply and return pipelines
3. Steam (S) boilers
4. High temperature water generators.
5. Heat Exchangers (HX)
6. Combustion Turbine Generator (CTG) (gas turbine)
7. Steam Turbine Generator (STG)
8. Condensers for steam turbines
9. Deaerators
10. HTW expansion tank (N_2 blanketed)
11. Electric motor-driven and steam turbine-driven chillers
12. Cooling towers
13. Thermal Energy Storage (TES) system
14. Major piping, pumps, and valves, and interconnections for subsystems

In the CP, different "types" of water are used, *e.g.*, high-temperature water (HTW), chilled water (CW), cooling tower water (CTW), steam (S), and steam condensate (SC). A control room provides an overview and status of the CP operation, which will be introduced in Chapter 6. In real-world operation, one wears proper clothing in power plants, including long pants, sturdy closed-toe and heel shoes, and long-sleeved shirts.

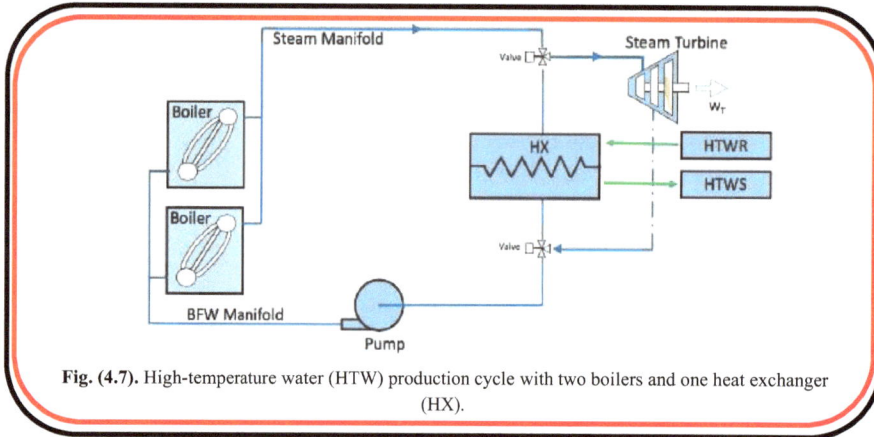

Fig. **(4.7)**. High-temperature water (HTW) production cycle with two boilers and one heat exchanger (HX).

4.6.1 High-temperature Water (HTW) Production Cycle

In the high-temperature water (HTW) production cycle, large HXs condense the steam to heat the high-temperature water return (HTWR) from campus and increase the HTWR temperature by about 150 °F. Fig. **(4.7)** illustrates the design of the system by considering two boilers and one heat exchanger, along with a steam turbine. In its production cycle:

a.) Fossil fuels, such as oils or natural gases, are injected and burned in boilers or a gas turbine. Steam is directly generated in the boilers. Note that there are two boilers equipped with the gas turbine to fully use the waste heat of its exhaust gas for steam production. A steam drum is located at the top of each boiler with a Liquid Level Gauge equipped to monitor the liquid level in the drum. A low level of liquid has a risk of overheating the internal boiling units. Note that as long as there is liquid water, the produced vapor in the steam drum remains saturated. The steam from the drum can be superheated by further raising its temperature using the boiler or additional heating source.

b.) The boiler steam is split into two directions: one is to the tube-shell heat exchangers (HX) where it transfers its thermal energy to the high-temperature water return (HTWR) from the campus. This increases the HTW temperature from 220 °F to 365 °F. Usually, the fluid in the shell side is subject to a phase change (*e.g.* vapor to liquid), releasing the latent heat to the fluid on the tube side. Using the HX and steam for heating HTW has a few advantages over direct combustion heating, such as the use of existing boilers for steam production and easy control/treatment of waters in the pipelines and boilers. The other direction of the stream is to steam turbines for either power generation or to a chiller compressor for chilled water production.

c.) For the steam going to HX, steam condensate will be collected after the HX and pumped back to the economizer-boiler system as the boiler feed water (BFW). The economizer is a specially designed HX equipped in the gas turbine system to recover the thermal energy in the exhaust stack gas flow right before being rejected to the ambient. Its primary role is to preheat the BFW for the boilers. As to the portion of steam going to turbines, the following two subsections will describe the relevant operation.

In addition, there are many applications of shell-and-tube HX in a power plant, sometimes with specific names, such as the condenser and evaporator of a chiller or steam turbine.

4.6.2. Chilled Water (CW) Production Cycle

In the chilled water (CW) production cycle, large chillers are operated to remove the thermal energy in the chilled water return (CWR) from campus and reduce its temperature by about 20 °F. The chillers are based on the mechanical compression vapor refrigeration cycle and require work input to drive compressors to enable heat transfer from low to high temperatures. Two types of compressor work are designed in the Central Plant: one is from electric motors powered by electricity generated by the Plant or imported from the Electric Utility Southern California Edison; the other is directly from the mechanical work of a steam turbine. Fig. (**4.5**) illustrates the CW production cycle design, along with a TES tank. In the CW production cycle:

a.) Mechanical work from the spinning shaft of a steam turbine or electric motor drives a compressor, which increases the refrigerant pressure and temperature and sends refrigerant to the condenser.

b.) In the condenser (essentially a HX), the refrigerant vapor becomes liquid due to the high pressure and the phase change releases the latent heat to the cooling tower water supply (CTWS) in the condenser's tube side.

c.) Liquid refrigerant passes through a throttling device which reduces the fluid pressure. The refrigerant then flows to the evaporator.

d.) In the evaporator, the liquid vaporizes due to the low pressure, absorbing thermal energy from the CWR in the tube side of the evaporator. This reduces the CWR temperature from about 50-60 °F to 40 °F. Then, the refrigerant vapor is back to (a).

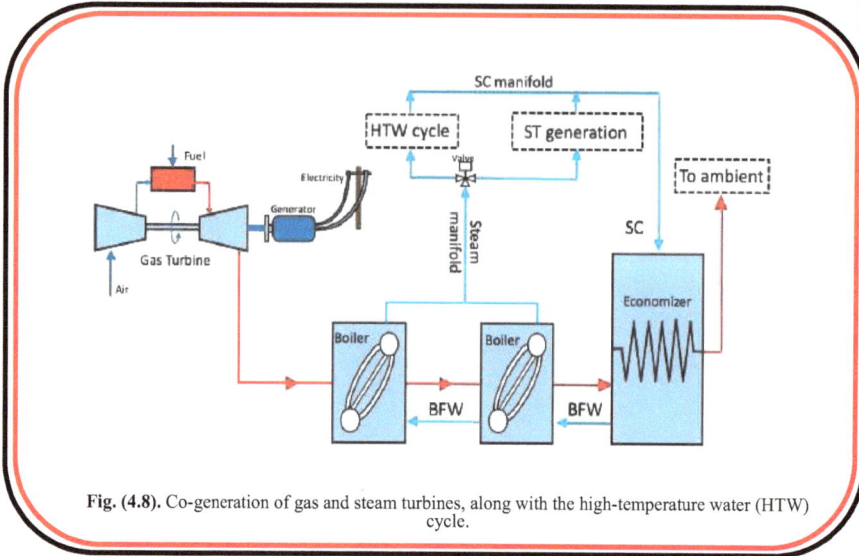

Fig. (4.8). Co-generation of gas and steam turbines, along with the high-temperature water (HTW) cycle.

4.6.3. Co-gen Electric Power Production

The UCI Plant has a steam turbine generator and combustion turbine generator. The combustion turbine exhausts to a heat recovery steam generator (HRSG), and the steam produced is used first to supply the high-temperature water (HTW) system; the remaining steam goes to the steam turbine generator or to the steam turbine chiller.

Fig. (**4.8**) illustrates the co-generation of gas and steam turbines and the high-temperature water (HTW) cycle. In operation, the steam turbine acts to regulate the steam system pressure and absorb any steam not used by the HTW system.

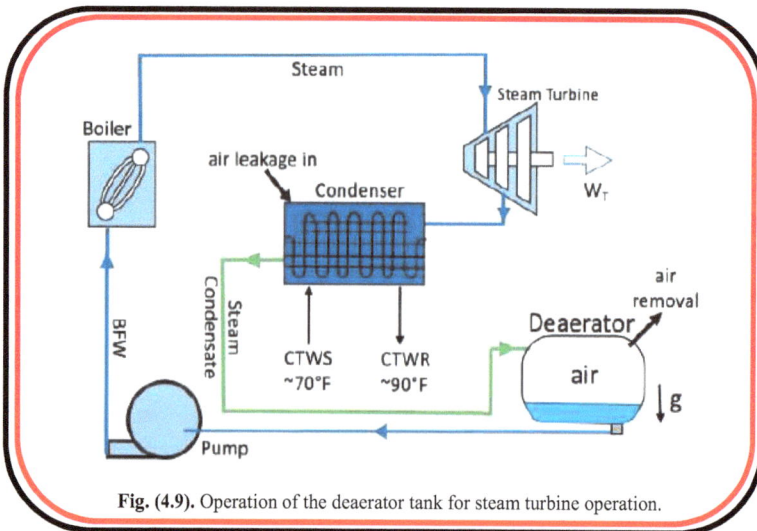

Fig. (4.9). Operation of the deaerator tank for steam turbine operation.

For the steam turbine,

a.) A portion of the steam (S) produced by boilers will be sent to a steam turbine for power generation at a scale of 0-5 MW.

b.) A condenser is equipped downstream of the steam turbine, where the steam condenses to liquid at a temperature around 90 °F by the cooling tower water supply (CTWS) in the tube side of the condenser.

c.) Liquid condensate flows through a deaerator tank to remove oxygen and other gases that leak into the system (in its condenser that operates at a vacuum) from the ambient and is then pumped back to the economizer-boiler system for steam production. Fig. (**4.9**) shows the deaerator connection and operation, which is to remove oxygen to reduce corrosion in the steel piping and tubes of the system.

For the gas turbine,

a.) Ambient air and fuel are injected into the gas turbine for power generation with the exhaust gas containing high-grade thermal energy at a temperature around 1,000 °F.

b.) The hot turbine exhaust gas flows through the heat recovery steam generator (HRSG), which has two boilers connected in parallel, and transfers its thermal energy to liquid water for high-pressure steam production in the boiler drums and to steam for superheated steam to supply the plant.

c.) Between the two boiler sections of the HRSG, there is a catalyst bed where CO reacts with O_2 to form more CO_2 and Ammonia (NH_3) reacts with the NO_x in the exhaust gases to reduce them to N_2 and H_2O to meet the stringent regulations of the South Coast Air Quality Management District as stated in the plant's Permit to Operate. These criteria emissions are monitored by the Continuous Emissions Monitoring System (CEMS).

d.) After the boiler section of the HRSG, the exhaust gas, still carrying a large amount of thermal energy, flows through an economizer installed downstream of the second boiler in the HRSG to preheat the boiler feedback water (BFW).

e.) Then, the exhaust gas is released to the ambient *via* the stack outlet pipe at a temperature of about 350 °F.

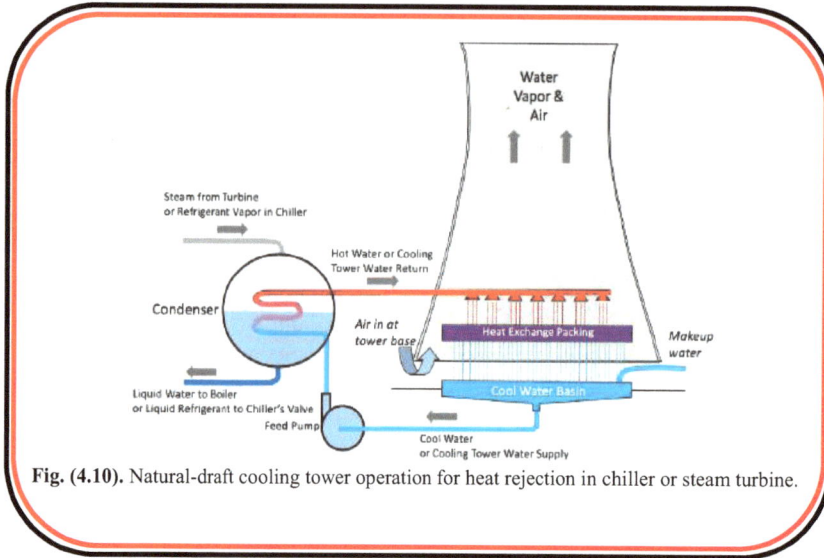

Fig. (4.10). Natural-draft cooling tower operation for heat rejection in chiller or steam turbine.

4.6.4. Cooling Tower Water (CTW) Cycle

Cooling towers provide the function of heat rejection to the ambient for CP operation, as shown in Fig. (**4.10**). Table **4.3** shows typical types of cooling towers in power plants. Two sets of cooling towers serve the two different sections of the CP.

The six Architectural Concrete Cooling Towers form a bank along the back of the Chiller and Boiler building. These towers are linked together at their sumps (though they can be isolated for service). The pumps from these towers discharge to a manifold (common header) and it is possible to serve any chiller from any tower or pump. In practice, the small differences in the head between pumping from one end to the other means using the pumps closest to the chillers. These towers and pumps cannot presently provide cooling tower water to the cogeneration plant.

The two fiberglass towers are housed behind the Cogeneration Building and serve only the condenser for the STG and lubricating oil coolers for the STG and GTG. These towers cannot, at present, serve the chillers.

The manifold arrangement of the pumps is widely adopted in power plant and engineering design, such as the steam manifold for boilers and BFW. In operation:

a.) Cooling tower water (CTW) of about 70 °F is supplied to the tube side of the condenser in either a steam turbine or chiller for heat removal, which is labeled as CTWS.

b.) In the condenser, the CTW temperature increases to about 80 °F and is then pumped back to the cooling towers for heat rejection to the ambient. The water is labeled as CTWR.

c.) CTWR is pumped to a higher level of the cooling tower and sprayed to small droplets, which fall by gravity. In the meanwhile, a fan above the sprayers is operated to draw ambient air, which is dry and cold, from the openings near the base and create a counter flow of air and water droplets. This counter flow inside the cooling tower thorough the mixing of water droplets and air, which improves heat exchange through forced convection, which evaporates a portion of the water.

d.) Droplets are collected at the tower bottom for CTWS of a temperature usually around 70 °F, which is then pumped to the condensers to close the cycle. At night, the ambient temperature is generally lower thus, the cooling towers are more efficient in heat rejection.

e.) On the air flow side, after absorbing heat and moisture from droplets, the dry cold air (*e.g.* about 70 °F) becomes humid and hotter (*e.g.* about 90 °F) and exits out of the cooling tower top. In cold weather, such humid air discharge encounters cold ambient above the cooling tower and condenses to tiny droplets, forming visible "smoke" or "fog." Because of CTW loss to ambient air, water is made up into the cooling tower sumps to maintain a constant level for the pumps to draw from.

#Example 2: In Figure 4.7, in a remodel design of the Central Plant to meet the increasing demand of the growing campus, the HTW production cycle needs to increase its capacity. Recommend economic plans.

Solution:

Plan#1: if the boiler is the limiting factor in the capacity increase, adding one or a few more boilers to the boiler manifolds in Figure 4.7, parallel with the existing boilers.

Plan#2: if the HX is the limiting factor in the capacity increase, adding one or a few more HXs to the HX manifolds in Figure 4.7 in parallel with the existing HXs.

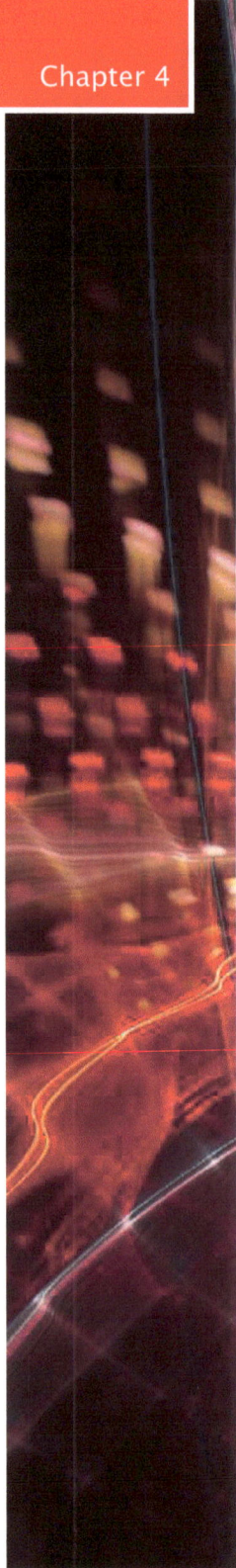

Table 4.3. Types of cooling towers.

Types	Function	Design
Natural-draft	Air enters the tower base and flows upward by buoyancy *via* a tall chimney. Water is sprayed to droplets to promote heat transfer between air and water.	
Mechanical-draft	Air flow is promoted by mechanical methods such as fans.	
Fan-assisted natural draft	A hybrid type that appears like a natural draft setup, while airflow is assisted by fans.	

4.6.5. Pipeline System & TES Tank

The Central Plant is connected to the HVAC systems of main buildings on campus *via* two major pipeline systems installed in a utility corridor ("tunnel") under the ring road of the campus, with the entrance located near the first chiller in the Plant.

    i.   HTWS and HTWR pipelines;

    ii.   CWS and CWR pipelines;

In operation, the plant supplies

a.) HTW of about 365 °F and CW of about 42 °F to the campus for heating and cooling, respectively. They are labeled HTWS and CWS.

b.) After heat exchange in the HAVC system of a building, HTW and CW return to the Plant at a temperature of about 210 °F and 50-60 °F, respectively. They are labeled as HTWR and CWR.

In the tunnel, the CW pipeline is much larger than the HTW one because HTW is about an order of magnitude greater than CW in the thermal energy delivery per mass or volume. This is due to the ratio of temperature differentials, about 150 °F for HTW and 10-20 °F for chilled water.

#Example 3: Assuming the campus needs equal heating and cooling loads, estimate the water flow rate ratio in the HWT and CW pipelines.

Solution:

From energy balance, the loading is expressed by:

$$\dot{Q} = \dot{m} C_v (T_S - T_R)$$

For HWT, $(T_S - T_R) \sim 150$ °F;

For CW, $(T_S - T_R) \sim 15$ °F.

Assuming the heat capacity C_v is approximately the same for CW and HTW,

$$\dot{m}_{HTW} / \dot{m}_{CW} \sim 1/10$$

The original design temperature differential for the plant chilled water system was 12 °F. By increasing the differential over the last 25 years, the plant has effectively doubled its pipeline carrying capacity. The HTW piping was designed for a much greater load than the plant experiences and is not near full capacity yet.

At design and normal operation, the HTW temperature is above the boiling point at ambient conditions; thus HTW must be pressurized to stay in its liquid state. In the UCI plant, this pressure is maintained by a nitrogen blanket tank in connection with the HTW pipeline. If the HTW pressure is lower than the design point, HTW will vaporize and lead to two-phase flow, which will damage pumps and cause water hammering in the pipelines. Fig. (**4.11**) shows the connection and operation of an N_2 pressurization tank. In operation of the nitrogen blanket tank:

a.) If the HTW pressure is higher than the design point, HTW will flow from the pipeline to the blanket tank, reducing the HTW pressure;
b.) If the HTW pressure drops below the design point, HTW will flow from the blanket tank to the pipeline, increasing the HTW pressure.

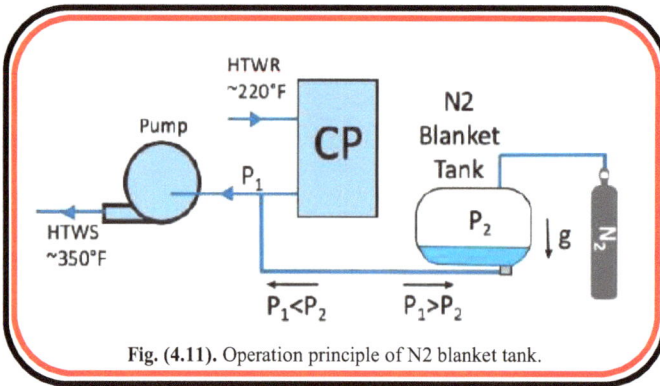

Fig. (**4.11**). Operation principle of N2 blanket tank.

In general, electricity is cheaper and ambient air is colder at night than in the daytime. Thus, chiller and cooling tower operation are more efficient and less costly at night. However, cooling is usually needed during the daytime when it is hot. To reduce cost and optimize operational efficiency, the University added a thermal energy storage tank (TES Tank) to the plant in 1997. The original design and method for operating the tank was as follows:

a.) CW at about 40 °F produced at night is pumped to the bottom of the TES tank;
b.) CW at about 40 °F in the TES tank is supplied to the campus for cooling in the daytime via the CWS pipeline;
c.) CWR of about 55 °F is pumped to the top level of the TES tank to be cooled by chillers at night.

Fig. (**4.5**) Shows the TES connection with the CW production unit. Gravity separates the CWR and CWS by the difference in density between waters at around 40 °F and 55 °F. No physical barrier exists in the tank.

#Example 4: The TES tank is a large cylinder 105 feet high and 88 feet in diameter; estimate the heat conductive flux and flow rate across the interface between CWR and CWS, assume the interface is a 1 m thick zone with linear temperature change. If CWS occupies half of the TES, compare heat loss at the interface in 24 hours with the "chilled" energy contained in CWS.

Solution:

h [m]	32.0
d [m]=2r	26.8
$dT=T_{CWR}-T_{CWS}$ [K]	6=283-277
A [m^2]=πr^2	565
dx [m]	1
$k \left[\dfrac{W}{mK}\right]$	0.584
$q_{cond} \left[\dfrac{W}{m^2}\right] = -\dfrac{kdT}{dx}$	**-3.24**
$\dot{Q}_{Loss}$ [W]= q_{cond} A	**-1833**
$C_V \left[\dfrac{J}{kg\,K}\right]$	4210
V [m^3]=$\dfrac{\pi r^2 h}{2}$	9041
$\rho \left[\dfrac{kg}{m^3}\right]$	1000
Q_{CWS} [J]= $\rho V C_V dT$	211000000000=211 [GJ]
Q_{Loss} @24 hrs [J]= $\dot{Q}_{Loss}t$	**-158364500=.1% Q_{CWS}**

With the addition of the co-gen plant in 2007, and 5 MW of solar photovoltaic panels in the early 2010's, the tank still works the same way. Now the time of creating chilled water changes with the availability of electricity, and one will see chillers operating during the day, which was nearly unheard of from 1997 to 2007. Operation of the plant is a continuous balancing act of thermal load, electric load, and cost for electricity. The skilled operators make decisions, minute-by-minute, to optimize operation.

For further readings regarding power generation and power plants, we refer the interested readers to the references in [5-11].

4.7. QUESTIONS

1. Is the steam generated in the boiler drum saturated or super-heated? How can you tell without instruments?

2. Describe the purpose of the cooling towers and list three mechanisms that enhance the purpose. When are they efficiently used? How do they work?

3. What will happen if the nitrogen-blanket tank pressure is too low?

4. Describe the purpose of the TES tank. Why does the CWR return from the top of the TES tank?

5. What is the purpose of a manifold? List three places in the Central Plant where a manifold is used.

6. Why is the HTW more efficient in thermal energy delivery than the CW?

7. In Co-gen, what is the benefit of using a steam turbine in gas turbine power generation?

8. Draw the schematic of the CP on one page with the major components properly connected.

9. In a chiller, what fluids are on the tube and shell sides of its condenser and evaporator?

10. Why using heat exchangers instead of directly heating the HTWR?

11. Describe the purpose of the pipeline system in the tunnel. Why are the CW pipes in the tunnel of a greater diameter than the HTW pipes?

12. "Smart meters" measure the use of electricity by households on a per hour basis. It has been proposed that electricity pricing should depend on the time of day it is used and should be most expensive at "peak hours" when average electricity consumption is at its highest. Why is it more expensive for electricity suppliers to generate electricity at these peak hours, and how

could "smart meters" be used to reduce average electricity costs for consumers?

13. What is the difference between power and energy?

14. Describe how you estimate the temperature of an unknown object if you don't have a thermometer with you.

15. Describe what world life looked like without power plants.

REFERENCES

[1]   I. G. Rice, U.S. Patent No. 3,703,807. Washington, DC: U.S. Patent and Trademark Office, 1972.

[2]   J. M. Baker, G. W. Clark, D. M. Harper, and L. O. Tomlinson, U.S. Patent No. 4,081,956. Washington, DC: U.S. Patent and Trademark Office, 1978.

[3]   M. Adolf, U.S. Patent No. 1,895,003. Washington, DC: U.S. Patent and Trademark Office, 1933.

[4]   C. E. Schutt, U.S. Patent No. 1,969,187. Washington, DC: U.S. Patent and Trademark Office, 1934.

[5]   A. W. Rankin, U.S. Patent No. 3,163,009. Washington, DC: U.S. Patent and Trademark Office, 1964.

[6]   R. E. Sonntag, C. Borgnakke, G. J. Van Wylen, and S. Van Wyk, Fundamentals of thermodynamics, New York: Wiley, 1998.

[7]   Y. A. Cengel and M.A. Boles, "Thermodynamics: An engineering approach", The McGraw-Hill Companies, Inc., New York, 2007.

[8]   R. Kehlhofer, F. Hannemann, B. Rukes and F. Stirnimann, "Combined-cycle gas & steam turbine power plants", Pennwell Books, 2009.

[9]   T. Tanuma, "Advances in steam turbines for modern power plants", Woodhead Publishing, 2017.

[10] A. S. Leĭzerovich, "Steam turbines for modern fossil-fuel power plants", The Fairmont Press, Inc., 2008.

[11] Y. Wang and K. S. Chen, "PEM fuel cells: thermal and water management fundamentals". Momentum Press, 2013.

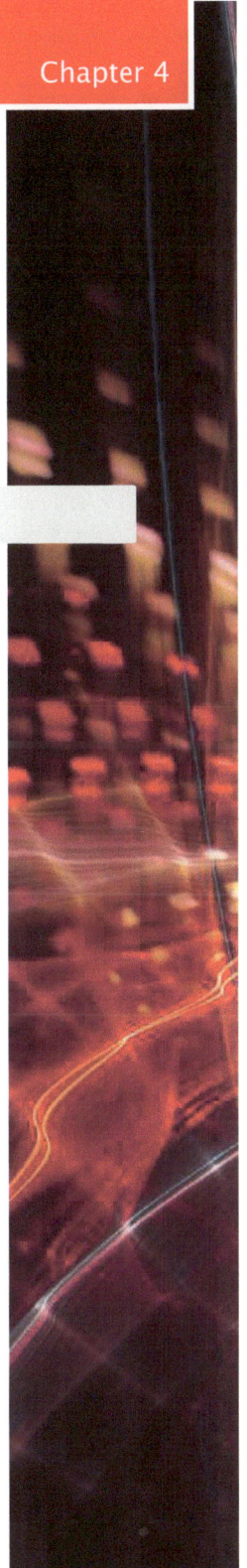

PIPE FLOW AND FLOW METERING

5.1. INTRODUCTION

Flow and volume measurement of incompressible fluids are prevalent in many engineering processes. Flows in a pipe network, also called pipe flow, are widely encountered in industry and our daily activities. For example, water is delivered to homes via the pipelines of a water distribution infrastructure. In power plants, numerous pipe flows are used to transport liquid, vapor, and two-phase flows: steam produced in boilers is transported in pipes to steam turbines for energy conversion. Boiler feed water (BFW) is pumped to a boiler via a BFW pipe.

In pipe system design, the main questions to be addressed are usually (1) what is the pipe diameter for a given flow rate, system length, and available pressure drop (2) for a given pipe and required flow rate, what is the pressure drop, and (3) what the optimum cost-effective design between pipe size and pumping energy use is. The flow rate and pressure drop determine the pumping power requirement, which is a major economic concern in system operations. Laminar flow in a straight section of a pipe or tube provides a classic example of the application of fluid mechanics equations to a practical problem.

Flow and volume meters are used in scientific work, industry, central energy plant operations, and infrastructure systems to quantify and measure flow rates for design, monitoring, or control purposes. Utilities use volume meters for revenue, such as the natural gas meter and the water meter serving your residence.

There are various types of flow rate and volume totalizing meter designs, with each usually having a specific accuracy, cost, and reliability. Among them, orifice plate and venturi meters remain common flow rate meters and measure pressure drop for the flow rate. Their operating principles are based on the basic fundamentals of fluid dynamics.

5.2. PRESSURE MEASUREMENT

Pressure is defined as a force over a unit of area. In fluids, several concepts are frequently used in pressure measurement, including absolute pressure, gauge pressure, and vacuum. The first is referred to as the absolute value of a force on a unit area of a wall that a fluid exerts. The second is the absolute pressure minus the local ambient (typically atmospheric) pressure. The third is that

pressure less than the local ambient. It can be described as the ambient minus the absolute, provided that the absolute does not exceed the ambient. In measurement, manometers are widely used, which measure liquid height and convert it to pressure using the hydraulic pressure imposed by the liquid height. A simple design is the U-tube manometers, in which the pressure difference between the two connected tubes is given by:

$$P = \rho g h \qquad\qquad \text{1}$$

where h is the difference of the liquid heights in the two tubes. An example of U-tube manometer is shown in Fig. (**S5.5**) [1].

A barometer is a device used to measure the local atmospheric pressure. A simple barometer follows the same principle as the U-tube manometer with mercury as the working fluid. Because the saturated vapor pressure of mercury is much lower than the ambient pressure under normal conditions, the vertical height of the mercury directly approximates the air pressure. Mercury's density is 13.6 kg/l, which leads to a 0.76 m liquid height under standard conditions. The Bourdon-tube pressure gauge is another widely used device, with an example shown in Fig. (**5.1**). It is based on the principle that a flattened tube tends to straighten or regain its circular form in the cross-section when pressurized. The elastic deformation causes displacement and transforms it to angular rotation of a pointer, indicating the pressure. Digital pressure transducers, often called pressure transmitters, convert pressure to analog electrical signals. They are widely used in scientific work for rapid high-accuracy measurement. Table **5.1** lists several major types of pressure measurement devices.

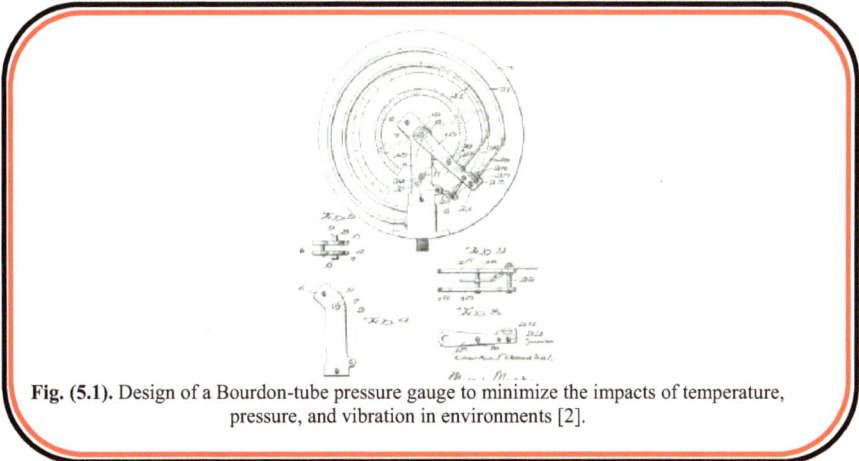

Fig. (**5.1**). Design of a Bourdon-tube pressure gauge to minimize the impacts of temperature, pressure, and vibration in environments [2].

Table 5.1. Types of pressure measurement instruments.

Types	Function	Image
Liquid Column Manometers	It consists of a column of liquid in a tube with the difference in liquid levels representing the pressure difference. It is widely used in ventilation, air conditioning, heating, dust elimination, *etc*.	
Bourdon-tube Pressure Gauge	It is based on the principle that a flattened tube tends to straighten or regain its circular form in the cross-section when pressurized. It measures gauge pressure and is widely used in aerospace, automotive, heating, *etc*.	
Digital Pressure Transducers	It uses an electrical circuit to convert the motion produced by a mechanical pressure element to electrical signals, which are measured to indicate pressure.	
Mechanical Displacement Type	It converts the pressure into mechanical displacement. Two common types are ring-balance and bell-type manometers.	

Table 5.2. Types of flow meters.

Types	Function	Image
Differential Pressure Flow Meter	Based on Bernoulli's equation and measure the pressure drop over an obstruction inserted in a flow. Widely used in industry.	
Positive Displacement Flow Meter	Use precision-fitted rotors as measurement elements. Known and fixed volumes are displaced between the rotors. The rotor rotation is proportional to the fluid volume being displaced. Work best with clean, non-corrosive/erosive liquids and gases.	

(Table 5.2) cont.....

Types	Function	Image
Mass Flow Meter	Measure the force induced by the flow for the flow rate. It can be designed to measure both liquid and gas flows.	
Open Channel Flow Meter	Measures the flow rate of an open channel by ultrasonic, microwave or radar technology using a non-contacting sensor mounted above the flume. They are widely used in water treatment plants, industrial waste applications, and irrigation systems.	

5.3. FLOW METER: VENTURI TUBE AND ORIFICE PLATE

Flow rate measurement is important to many engineering systems. Basic devices for flow rate/speed measurement range from lasers to delicate small heated wires to pressure drops across restrictions placed in the flow. This section will focus on two restrictive-type flow meters: the orifice plate and Venturi. They have no moving parts and produce pressure differences which are easily measured. Schematics of both meters are shown in Fig. (**5.2**).

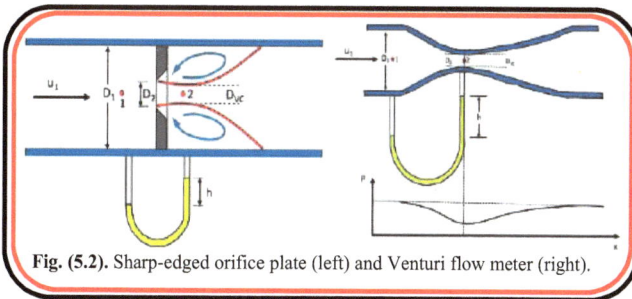

Fig. (5.2). Sharp-edged orifice plate (left) and Venturi flow meter (right).

Because the restrictions generally cause flow separation and associated turbulence, the flow devices are analyzed with an orifice or venturi coefficient as a function of geometry and flow Reynolds number. For example, a streamlined Venturi tube should approximately obey Bernoulli's equation:

$$\frac{p_1}{\rho} + \frac{u_1^2}{2} = \frac{p_2}{\rho} + \frac{u_2^2}{2} \qquad\qquad\qquad\qquad 2$$

between the approach flow (1) and the throat (2). With the areas of sections (1) and (2) denoted by A_1 and A_2, respectively, the mass balance equation will give:

$$A_1 u_1 = A_2 u_2 \text{ where } A_1/A_2 = (D_1/D_2)^2 \qquad\qquad 3$$

Combining Equations 2 and 3 will yield

$$u_2 = \sqrt{\frac{2(p_1-p_2)/\rho}{[1-\beta^4]}} \qquad\qquad\qquad\qquad 4$$

where $\beta = D_2/D_1$.

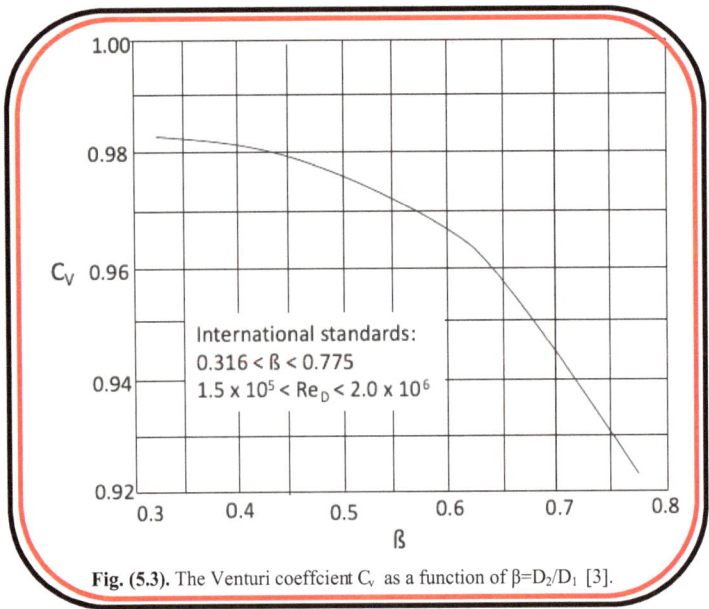

Fig. (5.3). The Venturi coeffcient C_v as a function of $\beta = D_2/D_1$ [3].

In operation, viscous forces will affect the value of u_2. Note that one assumption for Bernoulli's equation is inviscid flow. Thus, a coefficient is inserted in the particular geometry of the Venturi meter:

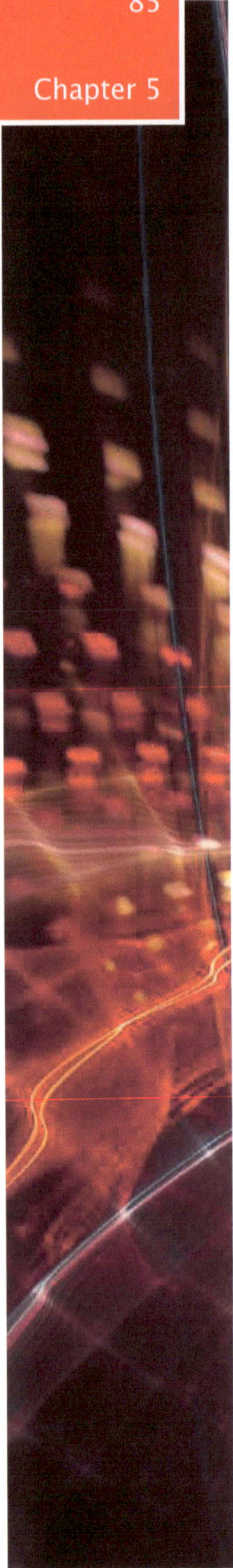

$$Q = C_v u_2 A_2 = C_v A_2 \sqrt{\frac{2(p_1-p_2)/\rho}{[1-\beta^4]}}$$ 5

where C_v is the Venturi coefficient and A_2 is the area of the throat. The profile of C_v is shown in Fig. (5.3), which typically varies from 0.94 to 0.98 and increases with the Reynolds number. In addition, an adjustable Venturi tube can be designed to fit various ranges of flow rate. An example is shown in Fig. (5.4).

 # Example 1: Calculate the Venturi coefficient C_v in the run#1 of Table 5.6.

Solution:

From the data of V and t, the flow rate is calculated by:

$$Q = \frac{V}{t} = \frac{745 \; ml}{3.17 \; s} = 235 \; ml/s \; or \; cm^3/s$$

The cross sectional area: $A_2 = \frac{1}{4}\pi D_2^2 = 0.785 \; cm^2$

The pressure drop: $p_1 - p_2 = \rho g \Delta h = 2{,}720 \; Pa$

From Equation 5,

$$C_v = \frac{Q}{A_2 \sqrt{\frac{2(p_1 - p_2)/\rho}{[1 - \beta^4]}}} = 1.25$$

An alternative to the elegant Venturi meter is the simple orifice plate flow meter. It is a plate with a small hole placed in the pipe run. The pressure drop across the plate is related to the flow rate. Although the flow rate derivation is similar to the Venturi, *i.e.*, based on Bernoulli's equation, flow separation turbulence downstream of the orifice causes large irreversible pressure loss. An orifice coefficient C_o is introduced in the same way as the Venturi coefficient C_v to account for non-ideal behaviors:

$$Q = C_o A_2 u_2 = C_o A_2 \sqrt{\frac{2(p_1-p_2)/\rho}{[1-\beta^4]}}$$ 6

where A_2 is taken to be the orifice area. In contrast to the Venturi meter coefficient C_v of almost unity, C_o is about 0.6, varying with the Reynolds number, size of the orifice relative to the pipe diameter and placement of the pressure taps (see Fig. (**S5.1**)). Two standard configurations of pressure tap placement are often used:

1) Radius taps, placed 2 radii upstream and one radius downstream;
2) Corner taps, placed in the meter flange immediately ahead and behind of the orifice plate.

Also, there are variations in the shape of the orifice – square-edged, chamfered, *etc*.

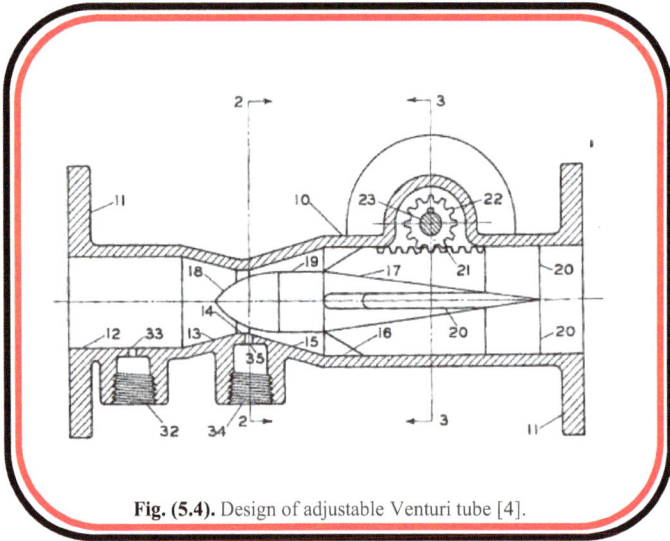

Fig. (5.4). Design of adjustable Venturi tube [4].

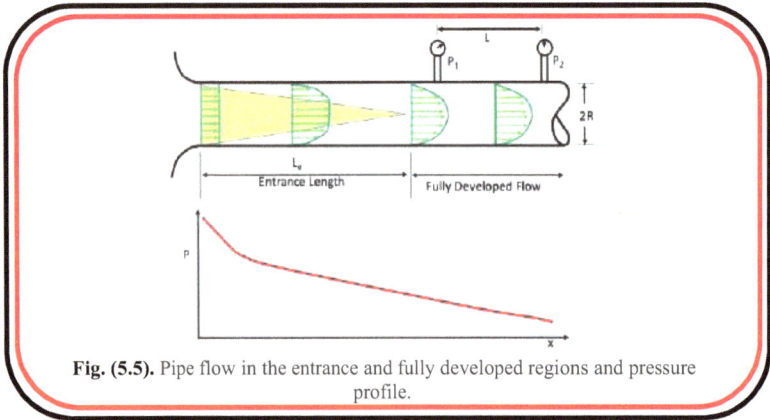

Fig. (5.5). Pipe flow in the entrance and fully developed regions and pressure profile.

5.4. PIPE FLOW

Steady, laminar flow driven by a pressure gradient in a straight pipe away from entrance effects is one of the few exact solutions of the Navie-Stokes equations in fluid mechanics. Two regions are generally defined: the entrance length and the fully developed region [5, 6], as shown in Fig. (**5.5**). The pressure acts normal to the cross-sectional area of the fluid in the pipe. A higher pressure gradient is present at the entrance length than at the latter region. In the fully developed region, the pressure difference $.p_1 - p_2)$ over the pipe length times the cross-sectional area (πR^2) results in a force in the direction of the flow:

$$F = (p_1 - p_2)\pi R^2 \qquad\qquad\qquad 7$$

where p_1 is the pressure at the upstream end, p_2 is the pressure at the downstream end, and R is the pipe radius, as shown in Fig. (**5.5**).

This force is balanced by the viscous shear stress force on the wetted area $(2\pi RL)$ of the pipe wall since at a steady state there is no acceleration of the fluid:

$$F = 2\pi RL\tau_w \qquad\qquad\qquad 8$$

where τ_w is the shear stress at the pipe wall and L is the length of the pipe between the pressure tap points p_1 and p_2. The pressure drop is $\Delta p = p_1 - p_2$ and the pressure gradient is $-\Delta p/L$. The minus sign arises due to the definition of the gradient as a function evaluated at point 2 minus the function at point 1, divided by the distance between the points.

In the force balance, equating these two forces yields:

$$\tau_w = \frac{1}{2}R\frac{p_1 - p_2}{L} \qquad\qquad\qquad 9$$

Since the force balance must also hold at any radius within the pipe, one will have:

$$\tau = \frac{1}{2}r\frac{p_1 - p_2}{L} \qquad\qquad\qquad 10$$

where τ is the shear stress at radius r. The shear stress varies linearly across the pipe from zero at the centerline to the maximum at the wall.

One can now invoke Newton's law of viscosity to relate the shear stress to the velocity gradient across the pipe:

$$\tau = -\mu \frac{du}{dr} \qquad \text{11}$$

Combination of the two preceding equations yields first-order ordinary differential equation (ODE):

$$\frac{du}{dr} = -\frac{1}{2\mu}\frac{p_1 - p_2}{L} r \qquad \text{12}$$

which integrates from the no-slip boundary condition at the pipe wall (u(r)=0 @ r = R) to r:

$$u(r) = \frac{R^2}{4\mu}\frac{p_1\text{-}p_2}{L}\left[1\text{-}\left(\frac{r}{R}\right)^2\right] \qquad \text{13}$$

This is the well-known Hagen-Poiseuille parabolic velocity profile for laminar flow in a pipe. The velocity profile is readily integrated to obtain the volumetric flow rate Q:

$$Q = \frac{\pi(p_1\text{-}p_2)R^4}{8\mu L} \qquad \text{14}$$

which relates the volumetric flow rate to the pressure drop. For laminar flow, the relationship is predicted to be linear.

 # Example 2: Calculate the flow rate Q using the pressure data in the run#1 of the 2.8 mm diameter pipes in Table 5.5, then compare with the flow rate data in the table. (L=360 mm and $\mu = 1.0 \times 10^{-3} Pa \cdot s$)

Solution:

The pressure drop is given by:

$$p_1 - p_2 = \rho g \Delta h = 1293 \ Pa$$

$R = \frac{2.8}{2} = 1.4 \ mm$

$$Q = \frac{\pi(p_1-p_2)R^4}{8\mu L} = 5.42 \times 10^{-6}\ m^3/s$$

In the experimental data,

$$Q_{exp} = \frac{V}{t} = \frac{44.0\ ml}{11.86\ s} = 3.71 \times 10^{-6}\ m^3/s < Q$$

Think about why they are different!

5.5. FRICTION FACTOR AND MOODY CHART

The above equation relating the flow rate to the pressure drop for laminar flow is dimensional. It is useful to cast the equation in dimensionless form. The Reynolds number is used to make the characteristic velocity in fluid flow dimensionless: it is the product of the characteristic velocity and a characteristic length scale divided by the kinematic viscosity. It represents the dimensionless ratio of inertial to viscous forces in a fluid. For pipe flow, the characteristic length scale is the diameter or radius; custom dictates the diameter D. Note that the characteristic length is not the length of the pipe since the inertial or viscous forces do not change along the length of the pipe. For the characteristic velocity scale, the average velocity $\bar{u}$ is used, given by the flow rate Q divided by the cross-sectional area. For laminar flow, it is determined by the pressure difference:

$$\bar{u} = \frac{(p_1-p_2)R^2}{8\mu L} \qquad \textbf{15}$$

The Reynolds number is then:

$$Re = \frac{\bar{u}D}{v} = \frac{\rho \bar{u} D}{\mu} \qquad \textbf{16}$$

where v is the kinematic viscosity v, ρ is the fluid density, and $D = 2R$.

Example 3: Calculate the Reynolds number using the flow rate data in the run#1 of the 2.8 mm diameter pipe in Table 5.5. (L=360 mm and $\mu = 1.0 \times 10^{-3} Pa \cdot s$)

Solution:

From the data and example 2,

$$Q_{exp} = \frac{V}{t}\frac{44.0\ ml}{11.86\ s} = 3.71 \times 10^{-6}\ m^3/s$$

The cross-sectional area: $A = \frac{1}{4}\pi D^2 = 6.15\ mm^2$

The average velocity $\bar{u} = \frac{Q_{exp}}{A} = 603\ mm/s = 0.603\ m/s$

From Equation 16,

$$Re = \frac{\rho \bar{u} D}{\mu} = 1{,}690$$

Though it is laminar flow, the pipe is very small comparing with those used in power plants. Thus, we usually face turbulent flows in the pipes of power plants.

Think about whether we can use the pressure data to calculate Re!

We seek a non-dimensional relation between the Reynolds number and a dimensionless force or pressure gradient that is exerted on the fluid. The pressure is made non-dimensional with the kinetic energy based on the average velocity. With some algebra, we find that:

$$\frac{\mu}{\rho \bar{u} D} = \frac{(p_1 - p_2)}{\frac{1}{2}\rho \bar{u}^2}\frac{D}{64L} \qquad \text{17}$$

The left side is $(Re)^{-1}$ and the right side is therefore dimensionless. We define the fanning friction factor f to be:

$$f = \frac{D}{L}\frac{(p_1 - p_2)}{\frac{1}{2}\rho \bar{u}^2} \qquad \text{18}$$

For laminar flow, this will yield:

$$f = \frac{64}{Re} \qquad \qquad 19$$

This is the dimensionless relationship between the pressure drop (f) and flow rate (Re) that was sought. For laminar flow, an exact relationship exists between f and Re.

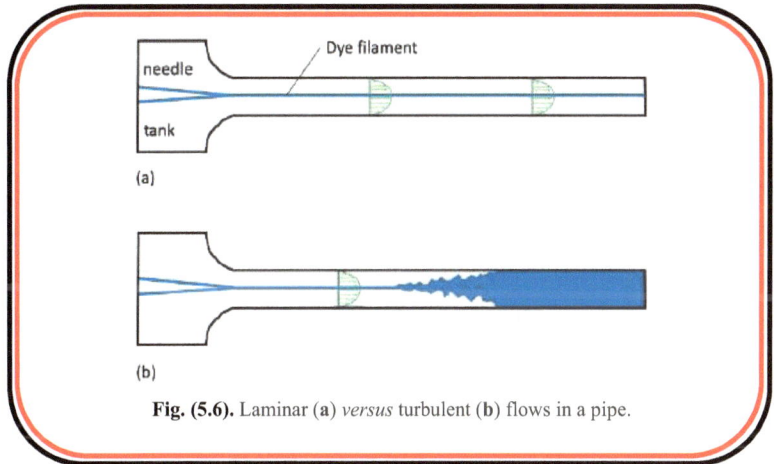

Fig. (5.6). Laminar (**a**) *versus* turbulent (**b**) flows in a pipe.

For turbulent flow, we have to rely on experiments. Note that there are other definitions in the engineering literature with different numerical coefficient. The Darcy friction factor is $\lambda = f/4$, so that $\lambda = 16/Re$ for laminar flow. Therefore, the user must be sure which definition is used. A friction-factor Reynolds number plot is shown in Figure S5.3 [3]. It spans the Reynolds number range 10^3-10^8 for smooth pipes and rough-walled pipes.

The laminar flow solution is a useful prelude to the more common case of turbulent flow in pipes. As the Reynolds number increases, the flow field becomes unstable with respect to small disturbances, as shown in Figure 5.6, say from a pump or vibration in the system. Above about Re = 2,300, most pipe flows are turbulent. Note that with extreme care, laminar pipe flow has been obtained to Reynolds numbers of about 90,000. While ubiquitous, turbulent flow is not amenable to an exact solution. The relationship between average shear stress and average velocity gradient is not known. The random velocity is decomposed into a mean part and a fluctuating part. We are interested in the mean flow properties, *e.g.* how is the volumetric flow rate of the mean velocity in the pipe related to the mean pressure drop. The relationship has been determined by experiment and semi-empirical theories. For the experimental determination of the pressure drop–flow rate characteristics in turbulent pipe flow, the friction factor – Reynolds number

framework still provides a useful aid. Since both are dimensionless, it means that we do not have to test different pipe diameters, lengths, fluid densities, and viscosities. We merely have to span the range of Re that will be found in applications and determine f. We could do the experiments with water or air, and the results should be the same on a f-Re plot. The plot could then be used to calculate the dimensional pressure drop for a specific design, say for gasoline in a pipe of known D and L for a specified flow rate. The only requirement is that the fluid is Newtonian.

One well-known semi-empirical formula for turbulent flow in smooth tubes is due to Blasius:

$$f = \frac{0.3164}{Re^{\frac{1}{4}}}, \qquad 2100 < Re < 100{,}000 \qquad\qquad\qquad \textbf{20}$$

It shows that f is higher for turbulent flow than for laminar flow extrapolated to Reynolds numbers higher than 2,300. Keep in mind that although f decreases as Re^{-1} in laminar flow and less in turbulent flow, the actual pressure drop increases with flow rate (linearly for laminar flow and non-linearly for turbulent flow) since p_1-p_2 is normalized with the average velocity squared. The pressure drop in turbulent flow is much larger than that if the laminar flow were extended to Reynolds numbers beyond 2,300.

 # Example 4: Calculate the fanning friction factor f in the run#1 for the 2.8 mm diameter pipe in Table 5.5.

Solution:
From Example 2,

$$p_1 - p_2 = \rho g \Delta h = 1{,}293 \; Pa$$

From Example 3,

$$\bar{u} = \frac{Q_{exp}}{A} = 0.603 \; m/s$$

From Equation 18,

$$f = \frac{D}{L}\frac{(p_1 - p_2)}{\frac{1}{2}\rho \bar{u}^2} = 0.0553$$

One can evaluate it from Equation 19 if pressure data are not available.

Table 5.3. Several friction factor correlations in different flow regimes. ϵ denotes the pipe wall roughness.

Laminar flow	$f = \dfrac{64}{Re}$
Smooth pipe turbulent flow	$f = \dfrac{0.316}{Re^{1/4}}$
Completely turbulent flow	$f = \left[1.14 + 2\log_{10}\left(\dfrac{D}{\epsilon}\right)\right]^{-2}$
Transition region	$f = \left\{-2\log_{10}\left[\dfrac{(\epsilon/D)}{3.7} + \dfrac{2.51}{Re(f^{1/2})}\right]\right\}^{-2}$

Since the friction factor – Reynolds number relationships are of a power-law type, log-log plots are used, as shown in Fig. (**S5.3**). In addition, roughness ϵ, as long as it is not too large, does not affect the laminar flow regime. However, it has a significant effect on turbulent pipe flow, as shown in Table **5.3**. As the height of the roughness elements increases, the friction factor does also, indicating more pressure drop is required for a given flow rate. The fundamental reason is the fluctuation of turbulent fluid flow in all directions. Fluctuation of moving fluid particles may send the particles to the cavities at the rough pipe surface, causing a momentum loss, as shown in Fig. (**S5.7**). Roughness height is usually expressed in empirical units rather than the actual height of the roughness elements since the shape and spacing of the individual elements affects the drag. Cast iron pipes were originally made in sand molds. The sand left an imprint on the cast pipe wall, and hence roughness is quoted in terms of equivalent sand-grain roughness. The limit of negligible roughness is called hydraulically smooth. The limit of large roughness is called fully rough, where the friction factor is large and independent of the Reynolds number. Since (p1-p2) is proportional to $\bar{u}$ in this case, it reflects the large pressure or drag across the roughness elements. Table **5.4** lists the absolute roughness of several pipe materials.

More correlations of the friction factor are given in Table **5.3** and a comprehensive picture of the friction factor is plotted in the Moody Chart in Fig. (**S5.3**) [3].

Table 5.4. Absolute roughness of several pipe materials [7].

Material	Roughness ϵ (in. x 10^{-3})	Roughness ϵ (μm)
Drawn tubing	0.06	1.5
Commercial steel	1.8	46
Wrought iron	1.8	46
Asphalted cast iron	4.7	120
Galvanized iron	5.9	150
Cast iron	10.2	260
Wood stave	6.5-32.7	180-900
Concrete	10.9-109.1	300-3000
Riveted steel	32.7-327.3	900-9000
Internally plastic coated	0.2×10^{-3}	5
Honed bare carbon steel	0.492×10^{-3}	12.5
Electropolished-bare Cr14	1.18×10^{-3}	30
Cement lining	1.3×10^{-3}	33
Bare carbon steel	1.38×10^{-3}	36
Fiberglass lining	1.5×10^{-3}	38
Bare Cr133	2.1×10^{-3}	55

5.6. EXPERIMENT

5.6.1. Apparatus

To experimentally verify the pipe flow theory and learn how to calibrate the two flow meters, an experiment example will be introduced in this section. A sample of testing data is given in Tables **5.5** and **5.6**. The apparatus consists of a flow bench with water as the testing fluid, as shown in Figs. (**S5.4 and S5.6**)

[3]. Different test pieces, such as pipes, meters, valves, *etc.*, can be inserted between a constant head source tank and an effluent tank whose elevation can be changed to vary the imposed pressure drop. Pressures are read from the two U-tube manometers that use the testing water itself as the manometer fluid. The flow rate is measured by timing flow into a graduated cylinder, as introduced in Chapter 2, instead owf a flow meter. The distance between the two tapping points for manometers is 360 mm. The pipe flow apparatus enables the measurements of both the flow rate (hence the Reynolds number) and pressure drop (hence the friction factor f), which will be used to compare with the Moody Chart in Fig. (**S5.3**) [3]. This apparatus also enables calibration of the restriction-type flow meters by determining the empirical coefficients. Samples of the calibrated empirical coefficients are shown in Fig. (**S5.1**) and Fig. (**5.3**).

Table 5.5. Pipe flow experimental data.

	2.8 mm Pipe (ε=0)				7.1 mm Pipe (ε=0)			
Run #	V (ml)	t (s)	P1 (mmH20)	P2 (mmH20)	Run #	V (ml)	t (s)	P1 (mmH20)	P2 (mmH20)
1	44.0	11.86	450	318	1	205	5.23	390	315
2	47.0	12.43	450	318	2	208	5.19	390	315
3	40.0	8.14	442	294	3	175	5.21	430	387
4	47.0	9.95	442	294	4	168	5.32	430	387
5	47.0	8.69	430	266	5	245	5.48	360	270
6	48.5	9.06	430	266	6	240	5.56	360	270
7	46.5	7.74	413	220	7	285	5.10	285	148
8	45.0	7.43	413	220	8	295	5.49	285	148
9	42.0	6.74	408	192	9	250	5.38	346	245
10	40.5	5.66	408	192	10	245	5.48	346	245
11	45.5	6.15	385	90	11	210	5.74	394	325
12	43.2	5.93	385	90	12	210	5.60	394	325

	10.6 mm Pipe (ε=0)							
Run #	V (ml)	t (s)	P1 (mmH20)	P2 (mmH20)				
1	480	5.13	38.5	34.0				
2	480	5.24	38.5	34.0		Parameter and Property		
3	525	5.12	35.2	29.2	L	360		mm
4	540	5.09	35.2	29.2	ρ	1000		kg/m
5	610	5.28	32.6	25.7	μ	8.90×10^{-4}		Pa·s
6	590	5.15	32.6	25.7				
7	650	5.12	29.3	21.0				
8	650	5.26	29.3	21.0				
9	685	5.10	26.2	17.1				
10	680	5.16	26.2	17.1				
11	805	5.28	19.0	7.1				
12	795	5.14	19.0	7.1				

5.6.2. Friction Factor (f)

a.) *Laminar Pipe Flow*

Laminar flow usually occurs under low Reynolds numbers (*e.g.* <2,300). The Reynolds number is evaluated by Equation 16 using the working fluid water viscosity. In the experiment, i) select proper pipe diameter and flow velocity to meet the Reynolds number requirement for laminar pipe flow. In general,

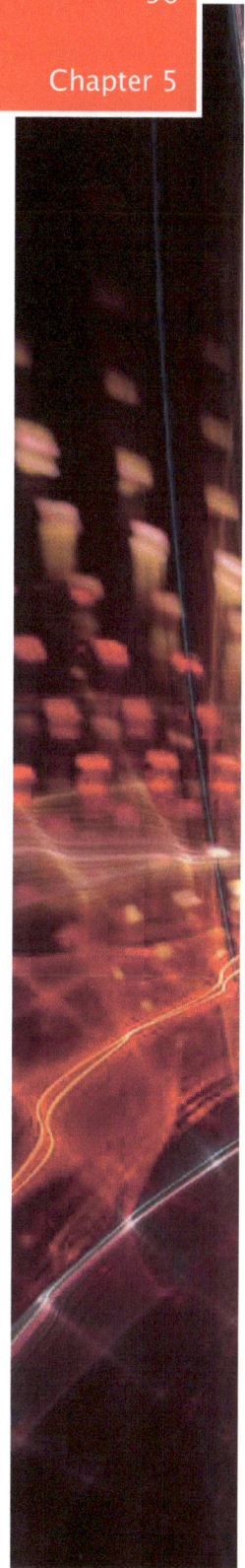

a sufficiently small pipe diameter (D) or slow flow rate will yield laminar flow. ii) connect the pipe to the flow bench and hook up the two manometers. iii) set a flow rate or velocity, which can be adjusted by changing the height of the pipe outlet, with a faster flow rate at a lower outlet height. iv) three measurements are taken:

- pressure drop using the manometer;
- liquid water volume in a graduated cylinder;
- duration to fill the volume using a stopwatch.

The pressure drop is used to calculate the friction factor f, while the volume and duration are for calculating the flow rate and the Reynolds number. v) change the outlet height for another flow rate.

b.) *Turbulent Pipe Flow*

Turbulent pipe flow occurs under high Reynolds numbers. In general, a sufficiently large pipe diameter (D) or fast flow rate will yield turbulent flow. Similarly, the Reynolds number needs to be evaluated prior to the experiment to ensure the pipe diameter yields the range of the Reynolds number of interest. The procedures and measurands are the same as the laminar flow.

Because turbulent flow occurs in a much wider range of the Reynolds number than laminar flow, additional pipe diameters can be tested. The experimental apparatus imposes an upper limit on the Reynolds number that can be tested in two aspects: 1.) the flow rate is limited by the maximum pressure drop of the apparatus; 2.) the upstream pressure tap of the manometer may fall in the entrance length of the pipe flow under high flow rate.

To measure the impact of the pipe wall roughness ϵ, different pipe materials can be tested, as shown in Table **5.4**. 3-D printed pipes can be used with a specific roughness design. One can carry out additional pipe testing of laminar flow under various roughnesses to verify its negligible impact on the friction factor. A sample of the experimental data using the above apparatus and procedure are given in Table **5.6** for exercise purpose.

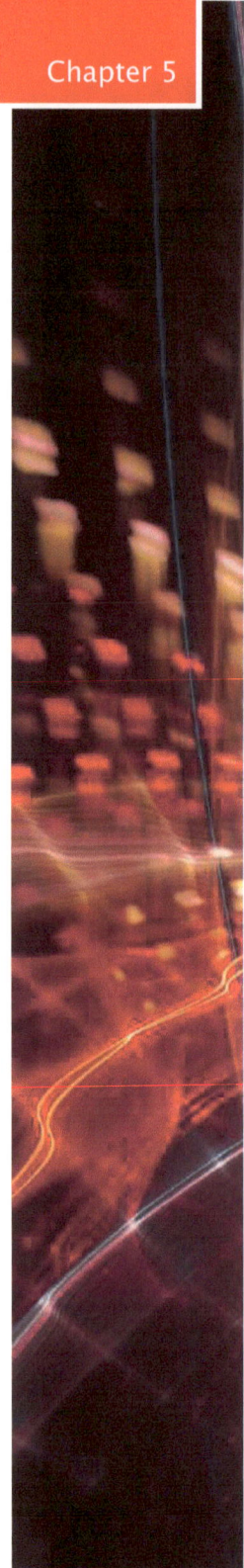

Table 5.6. Flow meter experimental data.

Venturi Tube (β=0.476 and D_2=10 mm)					Orifice Plate (β=0.545 and D_2=12 mm)				
Run #	V (ml)	t (s)	P1 (cmH2O)	P2 (cmH2O)	Run #	V (ml)	t (s)	P1 (cmH2O)	P2 (cmH20)
1	745	3.17	45.7	17.9	1	835	5.30	48.0	30.5
2	745	3.11	45.7	17.9	2	800	4.88	48.0	30.5
3	755	3.29	45.7	17.9	3	735	4.46	48.0	30.5
4	760	3.17	41.5	12.9	4	935	5.00	41.0	22.5
5	790	3.27	41.5	12.9	5	800	4.91	41.0	22.5
6	760	3.14	41.5	12.9	6	845	5.30	41.0	22.5
7	770	3.14	35.6	7.8	7	870	4.90	34.5	16.0
8	790	3.34	35.6	7.8	8	855	5.00	34.5	16.0
9	765	3.06	35.6	7.8	9	815	4.98	34.5	16.0
10	600	3.20	47.5	29.4	10	580	4.86	49.5	38.0
11	600	3.15	47.5	29.4	11	635	5.16	49.5	38.0
12	635	3.34	47.5	29.4	12	635	5.23	49.5	38.0

5.6.3. Flow Meter Calibration

Either Venturi or orifice plate flow metering requires calibration to determine the empirical coefficients, C_v and C_o, before use. A sample of C_v and C_o in calibration is plotted in Fig. (**S5.1**) and Fig. (**5.3**) with the configurations of the two flow meters given in Fig. (**S5.4**) [3]. Their calibration using the experimental apparatus is introduced in this section. A sample of the experimental data is given in Table **5.6**. In the experiment, i) insert a flow meter to the flow bench with all the taps hooked to the manometers, ii) set a flow rate, and iii) conduct three measurements:

- water heights (for pressure drop) in the manometers;
- liquid water volume in the graduated cylinder;
- duration to fill the volume using a stopwatch.

The pressure drop is used in Equation 5 or 6 for the empirical coefficient. Note that for the orifice, two pressure drops are measured for the radius and corner taps. One can evaluate their difference. The volume and duration are used to calculate the actual flow rate and the Reynolds number. The Reynolds number is calculated based on the up-stream tube diameter.

For further readings regarding fluid flows and experiments, we refer the interested readers to the references in [6-12].

5.7. QUESTIONS

1. In the experimental data sample, please estimate the precision of each equipment.

2. For pipe flow, calculate the friction factor, f, and Reynolds number for each pipe flow run. Tabulate the results and plot on the graph in Fig. (**S5.3**) [3] for comparison. Using engineering judgment, comment on comparison for agreement or lack of agreement.

3. For pipe flow, calculate the required pumping power in Watts for each pipe flow run. For 8 hours on-peak electric power ($0.25/kWh) and 16 hours off-peak power ($0.10/kWh) use per day, please calculate how much is required to run these flow for a day. Using engineering judgment, comment on the levels of power and cost.

4. Calculate the Venturi and orifice coefficients, and plot the results in Fig. (**S5.1**) [3] and Fig. (**5.3**). Using engineering judgment, comment on the comparison for agreement or lack of agreement.

5. Express the errors in f and Re as a function of the precisions of manometer, graduate cylinder, and stop watch in the pipe flow experiment. Note that the pressure and flow rate are independently measured.

6. For a Reynolds number of 100,000, calculate the difference in the friction factor f for smooth, fully rough, and laminar flow if it were somehow possible to maintain laminar flow to Re = 100,000. What is the ratio of the required pumping powers?

7. What are the advantages and disadvantages of flow-restriction meters such as the orifice plate and Venturi?

8. Why do we remove the air or air bubble in the manometer? If there is a 1.5 cm air length in the manometer pipe, estimate how much error it causes in the experimental results? Why?

9. Explain why pipe roughness only impacts turbulent flow.

10. Derive the pressure drops of a pipe flow as a function of diameter and maximum velocity, respectively, for the fully developed laminar flow.

11. Plot the f-Re relationships in Table **5.3** in the Re range of 1-1,000,000 for a smooth pipe and rough pipe of a relative roughness of 0.01.

12. If you were to write a patent of the Venturi used in the experiment, draw a schematic following Fig. (**5.1**) and briefly explain how it works.

13. How would you redesign an orifice plate flow meter to be 20% lighter?

14. How would you physically measure the flow rate of the Mississippi River?

15. Explain the relationship between velocity and the cross-sectional area as fluid flows through a pipe of changing cross-sectional area?

16. What causes users of a new device (*e.g.* orifice plate) to use it incorrectly? Give two possibilities.

REFERENCES

[1] D. A. Wozniak, U.S. Patent No. 4,380,173. Washington, DC: U.S. Patent and Trademark Office, 1983.

[2] H. C. J. Odend, U.S. Patent No. 2,027,875. Washington, DC: U.S. Patent and Trademark Office, 1936.

[3] MAE107 book. Available: https://tinyurl.com/MAE107book-figures

[4] J. M. Montgomery and P. Eric, U.S. Patent No. 2,240,119. Washington, DC: U.S. Patent and Trademark Office, 1941.

[5]   S. C. Cho, Y. Wang, and K. S. Chen, "Droplet dynamics in a polymer electrolyte fuel cell gas flow channel: Forces, deformation, and detachment. I: Theoretical and numerical analyses," , Journal of Power Sources, vol. 206, pp. 119-128, 2012.

[6]   F. M. White, "Fluid mechanics, 1999," Me Graw-Hill, 1979.

[7]   F. F. Farshad and H. H. Rieke, "Surface roughness design values for modern pipes". SPE Drilling & Completion, 21(03), 212-215, 2006.

[8]   L.F. Moody, "Friction factors for pipe flow. Trans. Asme, 66, 671-684, 1944.

[9]   J. Wu, and Y. Wang, "Liquid blockage and flow maldistribution of two-phase flow in two parallel thin micro-channels". Applied Thermal Engineering, 182, 116127, 2021.

[10] J. M. Lewis and Y. Wang, "Two-phase frictional pressure drop in a thin mixed-wettability microchannel". International Journal of Heat and Mass Transfer, 128, 649-667, 2019.

[11] S. C. Cho and Y. Wang, "Two-phase flow dynamics in a micro hydrophilic channel: A theoretical and experimental study". International Journal of Heat and Mass Transfer, 70, 340-352, 2014.

[12] X. C. Adroher and Y. Wang, "Ex situ and modeling study of two-phase flow in a single channel of polymer electrolyte membrane fuel cells". Journal of Power Sources, 196(22), 9544-9551, 2011.

EFFICIENCIES IN POWER PLANT

6.1. INTRODUCTION

In thermodynamics, several major devices related to energy conversion are introduced, including the internal combustion engines (ICE), steam turbine, boiler, heat exchanger, and heat pump. Thermal efficiency is a dimensionless performance measure of these devices. In general, thermal efficiency is the fraction of energy addition in the form of heat or thermal energy converted to useful output, as shown in Fig. (**6.1**), given as a percentage value. The nominally Otto-cycle ICE in automobiles can reach about 30% efficiency at the flywheel. Rankine-cycle steam turbine thermal efficiency can be as high as 41%. In heat pump cycles where heat rejection in the high-temperature side is the useful output, the efficiency is usually defined as the ratio of the rejected heat to the compressor work input, commonly called the coefficient of performance (COP). Refrigeration moves heat from a confined space and dissipates it in the atmosphere. Heat pumps move heat from one spot (often from the atmosphere or underground) to a home of business. Residential refrigerators and air conditioners are based on a vapor-compression mechanical refrigeration cycle. The former generally has a COP over 1 in practice, while the latter may have a COP over 3 to 5. The COP of heat pumps is generally higher than that of their refrigerator counterparts.

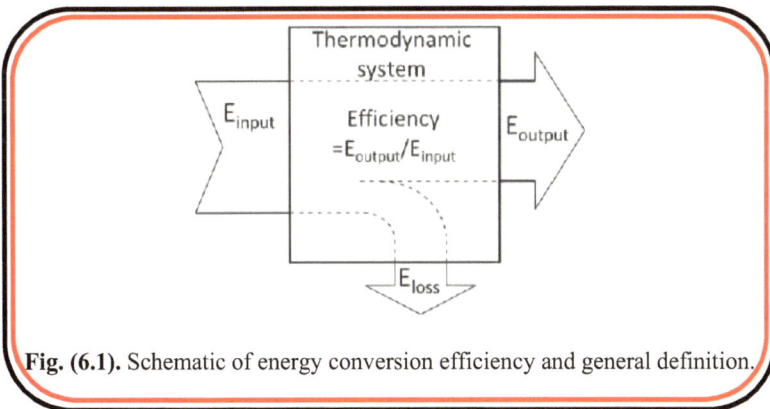

Fig. (6.1). Schematic of energy conversion efficiency and general definition.

In traditional power plants, the chemical energy in fossil fuels, such as coals and natural gases, is converted to electricity through heat engines, usually steam turbines. The efficiency is commonly expressed in terms of Heat Rate:

$$Heat\ Rate = \frac{Thermal\ Energy\ Input}{Electrical\ Energy\ Output} \qquad\qquad 1$$

Though the above definition is dimensionless, Heat Rate is often written as energy per energy, such as BTU/KWh and MJ/KWh. The average annual operating heat rate of the U.S. coal-fired power plants is approximately 10,400 Btu/kWh, according to the data in 2015 [1]. To convert to a percentage of efficiency, one can divide the equivalent Btu content of a kWh of electricity (3,412 Btu) by Heat Rate.

#Example 1: Calculate the power plant efficiency of a Heat Rate: (a) 10,500 Btu and (b) 7,500 Btu.

Solution:

(a) Efficiency $= \frac{3,412\ Btu}{Heat\ Rate} = \frac{3,412\ Btu}{10,500\ Btu} = 32.5\%;$

(b) Efficiency $= \frac{3,412\ Btu}{7,500\ Btu} = 45.5\%.$

Table **6.1** lists the heat rates of power plants using different energy resources.

Table 6.1 Approximate Heat Rates for electricity net generation [2].

Type of Power Plant	Approximate Heat Rate (BTU/kWh)
Coal	10,500
Petroleum	11,100
Natural Gas	7,800
Nuclear	10,500
Noncombustible Renewable Energy	9,100

6.2. EFFICIENCIES IN POWER PLANT

In a power plant, various components are installed to enable power generation, such as steam turbines, rotating electrical generators, boilers, cooling towers, and heat exchangers. Each of these components operates at a specific efficiency. The product of all of the different efficiencies is the overall plant efficiency.

a) Steam turbines are popular heat engines in power plants. They convert the thermal energy in steam to mechanical energy at an efficiency as high as about 40%.
b) The mechanical energy in the form of spinning movement of the shaft is converted to electricity by a rotating electrical generator at an efficiency that may be over 95%.
c) i.) In traditional power plants, boilers produce steam by converting the chemical energy in fossil fuels to heat through combustion and then use the heat to vaporize liquid water for steam turbines. Large central station coal-fired steam generators can convert more than 90% of the chemical energy in the coal to steam. Smaller natural-gas-fueled boilers are about 80% efficient.

ii.) In nuclear power plants, heat is generated by the nuclear reaction in a reactor, which converts the energy in nuclear fuels to steam production.

iii.) In centralized solar thermal power plants, the thermal energy source comes directly from the sun. High-grade thermal energy is generated by reflecting the solar beams by 100-1,000 mirrors to a single receiver at a reflectance above 90%. The thermal energy at the receiver is then transported and stored using molten salts. During the peak power hours, molten salts release heat to produce steam for steam turbine power generation.

To remove the heat rejected in steam turbine operation, a cooling tower system is installed to deliver the heat rejected by the condenser of a steam turbine via the cooling tower water to the ambient. The energy consumed for pumping cooling tower water and blowing air reduces the overall power-plant efficiency. In addition, water from rivers, lakes, or oceans is also commonly used for heat rejection.

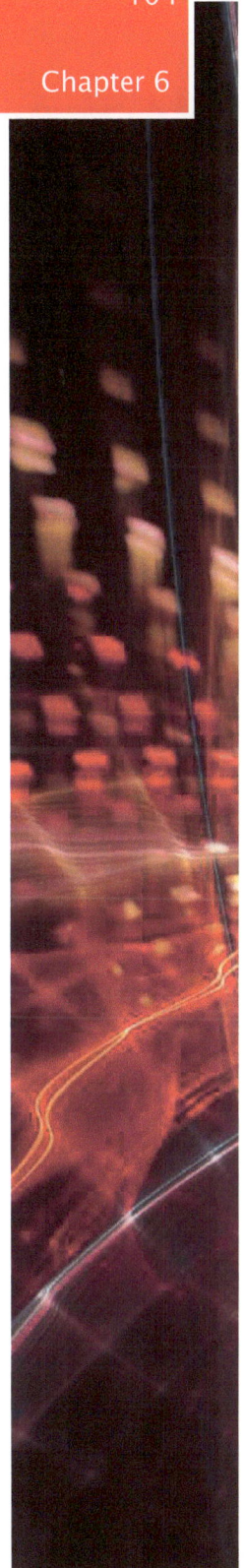

#*Example 2: Water Use in Power Generation: River water at 60 °F is used to remove heat from the condenser in the Rankine cycle of a power plant (100 MW). The power plant is at an efficiency of 31.5%. The cooling water returns to the river at 110 °F. (a) Estimate the rate of heat rejection in the condenser; (b) Calculate the flow rate of the cooling water.*

Solution:

a.) From the efficiency, the heat addition rate at the boiler side will be:

$$\dot{Q}_{add} = \frac{Power}{\eta} = \frac{100 \ MW}{31.5\%} = 317.5 \ MW$$

Thus, from the energy balance of the Rankine cycle,

$$\dot{Q}_{rej} = \dot{Q}_{add} - Power = 217.5 \ MW \ or \ 7.43 \times 10^8 Btu/hr$$

b.) From the energy balance of the cooling water,

$$\dot{Q}_{rej} = \dot{m}_{water} C_p (T_{out} - T_{in})$$

where the heat capacity C_p of water is 1.0 Btu/(lbm °F). Then,

$$\dot{m}_{water} = 1.49 \times 10^7 \ lbm/hr$$

6.3. OPEN SYSTEM ANALYSIS

In thermodynamic system analysis, three concepts are frequently used: open, isolated, and closed systems. An open system interacts with its surrounding by exchanging mass, heat, work, or other quantities at the boundary. In contrast, a system with no interactions of mass and energy with its surrounding is an isolated system. For a closed system, no mass is exchanged and energy is transferred at the boundary. For the major components in power plants, there are exchanges of both mass and energy. For example, the steam turbine-condenser system has steam flow in and condensate flows out with work production. In a gas turbine, air and fuel are injected with work and exhaust gas as outputs. In a boiler, water, air, and fuel are put in with exhaust gas and high-pressure steam flowing out. In a cooling tower, water returns and cold dry air flows in, and the cooling tower water supply and warm, humid air flows out. Fig. (**6.2**) shows a schematic of a steam turbine open system. For an overall system of the boiler, steam turbine, condenser, and pump, water (both liquid and vapor) can be treated as a closed system.

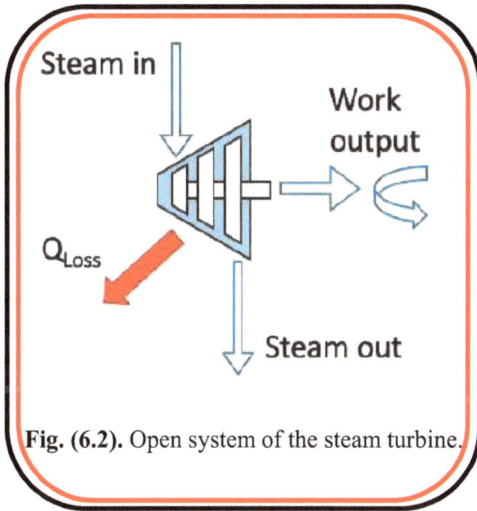

Fig. (6.2). Open system of the steam turbine.

6.4. BOILER AND BOILER IN CO-GENERATION

In boilers, steam (S) is produced by burning fossil fuels. Table **6.2** lists several typical boilers in the industry. In operation, boiler feed water (BFW) is pumped to boilers and is vaporized by the heat from fuel combustion. In commercial boilers, combustion, heat transfer, and water evaporation are optimized for steam production. Heat loss occurs at the boiler surface, which is minimized by using insulation materials. Fig. (**6.3**) shows the open system of a typical boiler. Boiler efficiency measures how effectively the thermal energy in fuel combustion is used to produce steam:

$$Boiler\ Efficiency = \frac{Energy\ for\ Steam\ Production}{Thermal\ Energy\ Input\ from\ fuel} \longrightarrow \quad 2$$

Table 6.2 Types of boilers in industry.

Types	Function	Figure
Fire Tube Boiler	Flames and hot gases pass through tubes running through a sealed container of water.	
Water Tube Boiler	Water circulates in tubes and is heated externally by flames and hot gases surrounding the tubes.	
Fluidized Bed Boiler	Fuel is burned within a hot bed of inert particles. The fuel-particle mix is suspended by air flowing upward within the bed.	
Stoker Fired Boiler	Boiler furnace is fed by a mechanical stoker.	
Pulverized Fuel Boiler	Steam is generated by burning pulverized fuel that is blown into the firebox.	
Waste Heat Boiler	Recover heat from the exhaust gases of other processes to generate steam.	

In energy analysis, the chemical energy in fuel can be evaluated by using the concept of the lower heating value (LHV), defined as the amount of heat released by combusting a unit mass/volume of fuel initially at 25 °C and returning the temperature of the combustion product to 150 °C. Water product exists in the vapor phase. As example, the LHV of natural gas (NG) is about 1,050 Btu/scf. In contrast, the higher heating value (HHV) adds the latent heat of water condensation, h_{vl}, under the same unit:

$$HHV = LHV + h_{vl} \qquad \qquad 3$$

Although in boilers, exhaust gas is generally over 150°C, and thus LHV would seem to make the most sense, a long-standing industry tradition is to use the higher heating value (HHV). Fuel is generally bought and sold based on the HHV, and thus use of the HHV for overall plant heat rate simplifies costs of operation estimates. Occasionally, one sees the LHV used. Table **6.3** lists the LHV and HHV values for a few fuels. By using the LHV, Equation 2 becomes:

$$Boiler\ Efficiency = \frac{Energy\ Rate\ for\ Steam\ Production}{LHV * m_{fuel}} \qquad 4$$

In the steam output, the energy rate can be approximated by the change in enthalpy:

$$Boiler\ Efficiency = \frac{h_s m_S - h_{BFW} m_{BFW}}{LHV * m_{fuel}} \qquad 5$$

Assuming no leakage, the mass balance will give:

$$m_S = m_{BFW} \qquad \qquad 6$$

Note that h_s is readily obtained from the steam table. If BFW is saturated water, then h_{BFW} is also available in the steam table. Otherwise, h_{BFW} can be obtained using the specific heat of liquid water and the temperature difference with the saturated liquid water.

Table 6.3. LHV and HHV of a few fuels [3.4].

Fuel	HHV [MJ/kg]	LHV [MJ/kg]
Hydrogen	141.80	119.96
Methane	55.50	50.00
Diesel	45.6	42.6
Gasoline	46.4	43.4
Ethanol	29.7	26.7

In co-generation that uses the high-grade thermal energy in the exhaust gas of a gas turbine for steam production, *e.g.* a waste heat boiler in Table **6.1**, efficiency expression needs to be modified because no fuel is combusted in the boiler. The boiler functions like a heat exchanger that involves the phase change of liquid water to steam. A formula can be developed by using the thermal energy input in the exhaust gas:

$$Boiler\ Efficiency = \frac{h_s \dot{m}_S - h_{BFW} \dot{m}_{BFW}}{(h_{in} - h_{out}) \dot{m}_{exhaust\ gas}} \qquad 7$$

In analysis, one can treat the exhaust gas as pure air using the Air Standard Assumption (to be explained in detail in Chapter 8):

$$Boiler\ Efficiency = \frac{h_s \dot{m}_S - h_{BFW} \dot{m}_{BFW}}{(h_{air,in} - h_{air,out}) \dot{m}_{exhaust\ gas}} \qquad 8$$

The air enthalpy is a function of temperature, expressed by:

$$h(T) = h_{T_r} + \int_{T_r}^{T} C_p(T) dT \qquad 9$$

where T_r denotes the reference temperature. A further simplification is to apply the Cold Air Standard Assumption that the specific heat is assumed constant and the same as that at 77 °F, then,

$$Boiler\ Efficiency = \frac{h_s \dot{m}_S - h_{BFW} \dot{m}_{BFW}}{C_{p,77F}(T_{in} - T_{out}) \dot{m}_{exhaust\ gas}} \qquad 10$$

Fig. (6.3). Open system of a typical boiler.

6.5. HEAT EXCHANGER EFFICIENCY

In heat exchangers (HX), heat is transferred between two or more fluids at their interfacial boundary or wall. Table **6.4** lists several popular types of HXs. In most cases, fluids are not mixed inside a HX. Commercial HXs usually create a large interfacial area by adding fins or other surface structures to maximize heat exchange efficiency. Heat loss occurs at the HX surface to the surrounding environment, which can be mitigated by using insulation materials. Fig. (**6.4**) shows a HX open system. The HX efficiency can be defined as the ratio of the actual heat transfer rate to the maximum possible heat transfer rate:

$$\text{HX Efficiency} = \frac{\text{Actual Heat Transfer Rate}}{\text{Maximum Possible Heat Transfer Rate}} \qquad 11$$

Table 6.4 Types of Heat Exchangers.

Types	Function	Figure
Shell and tube heat exchanger	Composed of a shell containing a series of tubes. Fluids in the shell and tube bundle exchange heat.	

(Table 6.4) cont.....

Types	Function	Figure
Plate heat exchanger	Composed of a number of thin, slightly separated metal plates that have large surface areas and small fluid flow passages for heat transfer between two fluids.	
Plate and shell heat exchanger	Combination of plate and shell-tube heat exchangers	
Plate fin heat exchanger	Plates and finned chambers are employed to transfer heat between fluids.	
Pillow plate heat exchanger	Wavy pillow-shaped plate geometry. A heating or cooling medium circulates in the space between the plates to transfer heat.	
Phase-change heat exchanger	To heat a liquid for evaporation or boiling like evaporators or boilers or to cool a vapor for condensation like condensers.	

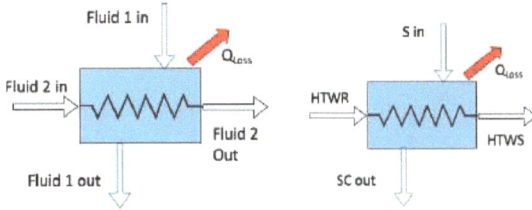

Fig. (6.4). (Left) Open system of a heat exchanger (HX); (Right) Open system of the HX in the Central Plant where steam (S) and steam condensate (SC) are in the shell side and high temperature water return (HTWR) and supply (HTWS) are in the tube side.

For an HX that uses steam (S) as Fluid 1 and liquid water as Fluid 2 as shown in Fig. (6.4) without mixing, the HX efficiency can be written as:

$$\text{HX Efficiency} = \frac{\dot{m}_{w,2} C_P (T_{out,2} - T_{in,2})}{\dot{m}_{S,1}(h_{s,in,1} - h_{sc,out,1})} \qquad\qquad 12$$

#*Example 3: For a heat exchange of Figure 6.4 (right), $\dot{V}_{HTW}$=210 gpm, $\dot{m}_S$ =28,000 pph, P_s=220 Psig, P_{sc}=150 Psig, T_{HTWS}=365 °F and T_{HTWR}=210 °F. Calculate $\dot{Q}_{Loss}$.*

Solution:

With the data provided, we can get the enthalpy values from Steam tables and interpolating them to get:

$P_s = 220\ Psig + 15 = 235 Psia$	$h_s = 1200.76 \dfrac{Btu}{lbm}$
$P_{sc} = 150\ Psig + 15 = 165 Psia$	$h_{sc} = 338.60 \dfrac{Btu}{lbm}$

For $\dot{m}_{HTW}$:

$$\dot{m}_{HTW} = \dot{V}_{HTW}\rho_{HTW} = 12600 \left[\frac{gal}{hr}\right] * 7.712 \left[\frac{lbm}{gal}\right] = 97171 \left[\frac{lbm}{hr}\right]$$

From energy balance,

$$Q_{loss} = \dot{m}_S(h_S - h_{SC}) - \dot{m}_{HTW} C_P(T_{HTWS} - T_{HTWR})$$

where the water heat capacity C_P is 1.0 $\left[\frac{Btu}{lbm\ °F}\right]$, thus:

$$Q_{loss} = 9 \left[\frac{MMBTU}{hr}\right]$$

6.6. STEAM/GAS TURBINE EFFICIENCY

In a steam turbine (ST), high-pressure steam is fed in with its thermal energy converted to mechanical work. An ST open system is shown in Fig. (**6.2**). The ST efficiency can be expressed by:

$$\text{ST Efficiency} = \frac{\dot{W}_{ST}}{(h_{s,in} - h_{s,out})\dot{m}_S} \qquad \qquad 13$$

In a gas turbine (GT), fuel and air are injected for combustion, with the associated thermal energy converted to work. The GT efficiency can be expressed by:

$$\text{GT Efficiency} = \frac{\dot{W}_{GT}}{LHV \times \dot{m}_{fuel}} \qquad \qquad 14$$

Note the LHV is used for the efficiency calculation of gas turbines (also commonly called combustion turbines) in the above equation. Developed more than a century after the earliest steam systems, the gas turbine industry developed different traditions. For electric power generation, the spinning shaft of ST or GT is connected with a rotary electric generator for electricity production. The electric power can be used to replace $\dot{W}$ in the above two equations, in which the efficiency then represents the overall efficiency of steam (STG) or gas turbine generation (GTG).

6.7. POWER PLANT MONITORING AND DATA COLLECTION

In a power plant, components are monitored regularly with real-time data collection for control and diagnostics purposes. For example, the Central Plant at UC Irvine will be introduced in this section to show real-time data monitoring and efficiency evaluation of main devices and overall power generation using the real-time field data. Specifically, this section is to design procedures and collect appropriate data to determine the efficiencies of:

a. Boilers
b. Heat exchangers (HX)
c. STG
d. GTG
e. Co-generation.

Real-time field data are usually shown in the control room of a power plant using monitor screens. Samples of the data monitoring screens at the UC Irvine are shown in Figs. (**6.5**, **6.6**, and **6.7**) and Figs. (**S6.1** and **S6.2**) for different components. Additional or back-up measurements can be taken from gauges on equipment in the Plant.

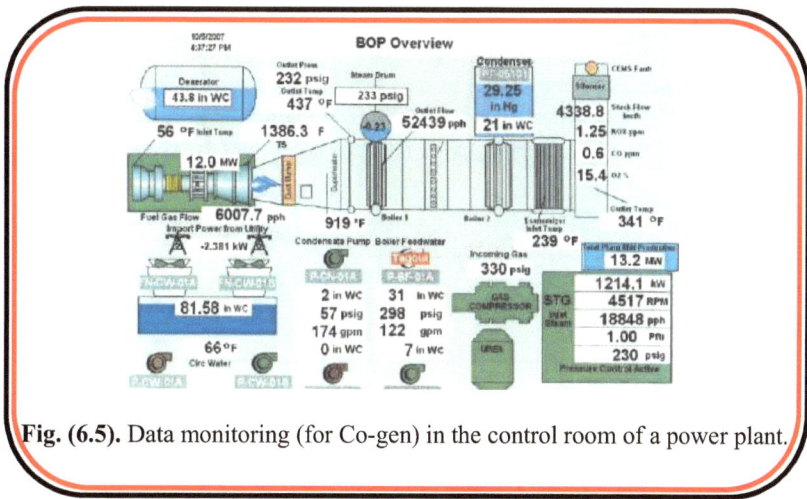

Fig. (6.5). Data monitoring (for Co-gen) in the control room of a power plant.

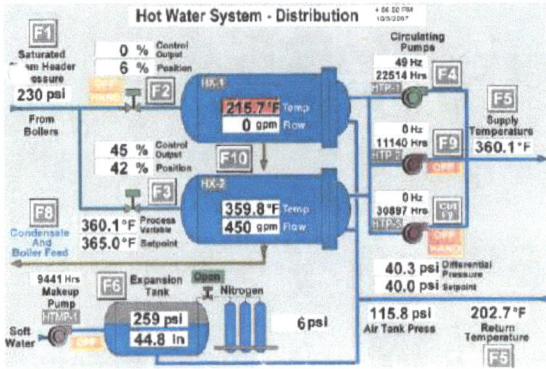

Fig. (6.6). Data monitoring (for HXs) in the control room of a power plant.

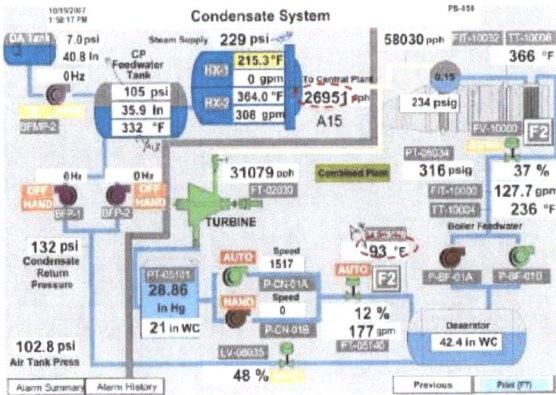

Fig. (6.7). Data monitoring (for HXs+STG) in the control room of a power plant.

In traditional field data collection, a proper table format should be prepared to list parameters to be measured, as shown in Fig. (**6.8**). Because nowadays cameras are available in cell phones, it is strongly suggested to take pictures of the screens in case the data needs to be rechecked later. Then, all the data are taken out and tabulated professionally.

Fig. (6.8). Sample of table format for field data recording.

In addition, certain field data, as shown in the control room, may not be reliable due to multiple reasons, such as sensor malfunctions and the expiration of services. For data that are unimportant in plant operation, they may not be accurate after many years of operation. For example, the stack exhaust gas flow rate is found to be inaccurate in the Plant. To obtain a more accurate value for the stack exhaust gas flow rate, one can calculate the mass balance from other reliable data such as the fuel injection rate and oxygen fraction in the exhaust gas flow. Note that the fuel injection rate is directly related to the operational cost that a power plant pays for electricity generation. The oxygen fraction, which can be accurately sensed, is important to monitor the combustion stoichiometric ratio and ensure lean combustion and full fuel utilization. Assuming the volume ratio of O_2 to N_2 is 1:4 in the ambient air and CH_4 is the main component of natural gas (NG) fuel, the chemical reaction formula can be developed as below:

$$CH_4 + (x+2)O_2 + (4x+8)N_2 \rightarrow CO_2 + 2H_2O + (4x+8)N_2 + (x)O_2 \qquad \textbf{15}$$

The right side is the composition of the exhaust gas, which gives the content of O_2:

$$O_2\% = \frac{x}{1+2+(4x+8)+x} = \frac{x}{11+5x} \qquad \textbf{16}$$

With the field data of $O_2\%$, one can calculate x and then the mass flow rates of the ambient air and exhaust gas from the fuel injection rate using Equation 15.

#Example 4: O_2% in the exhaust pipe is 15%. Calculate $\dot{m}_{CO_2}$ and $\dot{m}_{air}$ for a CH_4 injection rate of 6573 pph.

Solution:

Knowing we have 15% of O_2, from equation 16 one will reach:
$$.15(11 + 5x) = x$$
→*$x = 6.6$, then the chemical equation is written by:*

$$CH_4 + 8.6O_2 + 34.4N_2 \rightarrow CO_2 + H_2O + 34.4N_2 + 6.6O_2$$

Knowing we have an injection of $6573\frac{lb}{h}$ of fuel:

$$\dot{m}_{CO_2} = \dot{m}_{CH_4}\left(\frac{M_{CO_2}}{M_{CH_4}}\right) = 18,035 \left[\frac{lb}{hr}\right]$$

The mass flow rate of air is given by:

$$\dot{m}_{air} = \left(\frac{(x+2)M_{O_2} + (4x+8)M_{N_2}}{M_{CH_4}}\right)(\dot{m}_{CH_4}) = 697,946 \left[\frac{lb}{hr}\right]$$

For further readings regarding power plant or power generation efficiencies, we refer the interested readers to the references in [5-9].

6.8. QUESTIONS

1. Draw the (1) boiler(s)-economizer (as one system), (2) heat exchanger(s), (3) STG, and (4) boilers (including economizer)-HX-STG open system block diagrams showing main inputs, outputs and other relevant information (main control valves, *etc.*). Clearly label all flow streams.

2. Formulate the open system balance equations for each of the above and define their efficiencies. From the tabulated data, plot the relevant observed data *versus* time. Group similar data into the same plot when appropriate.

3. Formulate and calculate the heating load to the campus (the rate of heat delivered by HTW) from the data in a spreadsheet and plot. Use units that make sense, *e.g.*, MMBtu/hr.

5. Formulate and estimate the CO_2 production and air injection rate in the co-generation and plot. Use units that make sense, *e.g.*, pounds mass/hour. Since CO_2 is one of the greenhouse gases, comment on the impact of the Central Plant on global warming based on the estimate (Assume CH_4 is the only component in the natural gas).

6. For the boiler(s) (including the economizer), calculate and plot the components of the open system balance formula, *i.e.*, "input" from natural or exhaust gas, output of the steam, and heat loss in consistent units, *versus* time.

7. Calculate and plot the efficiencies of the boilers (including the economizer), heat exchanger, STG, and GTG *versus* time.

8. For the boiler-HX system (assuming no STG), use the pressure data of the high pressure S from boiler, SC after the HX and BFW pressure to sketch the P-v diagram for steam and show the path of the water/steam through the boiler.

9. Formulate an expression to calculate the percentage of enthalpy change of the water in the boiler system due to sensible heating relative to the total change, and comment on the values.

10. Use an expanded section of the attached steam $P - h$ diagram to show the thermodynamic path of the steam from the boiler through the control valve into the shell side of the heat exchanger(s). Clearly describe the change in the state of the steam in this process.

11. If the price for natural gas is $0.50 per "therm," (a "therm" is 100,000 Btu) and electricity from the outside grid is $0.11 per kWh. Can you formulate an expression to estimate how much the Central Plant saves per kWh by using Co-gen (STG+GTG) (if neglecting the HTW production) using the collected field data?

12. If a moving body has zero acceleration based upon F=ma, is the body still moving? Explain. Estimate the initial energy it has if initially moving at a speed of U and a temperature of T.

13. List three methods to improve the efficiency of a steam power plant.

14. List three methods to improve the boiler-economizer efficiency.

15. List three methods to improve the HX efficiency.

16. Calculate the efficiency for each power plant in Table **6.1**.

17. How much cooling for the nuclear reactor and cooling tower are required for the "Little Boy" if used for power generation for one year with a Heat Rate given in Table 6.1. Below are some facts about "Little Boy":
"On August 6, 1945, the United States dropped its first atomic bomb from a B-29 bomber plane called the Enola Gay over the city of Hiroshima, Japan. The "Little Boy" exploded with about 13 kilotons of force, leveling five square miles of the city and killing 80,000 people instantly."
[www.history.com]

18. Design and draw a "Little Boy" nuclear power plant (see the above) using two steam turbines in parallel, with the waste heat for residential heating using HTW, hydrogen production using the high-temperature electrolyzer technology, and pool heating for the city of Hiroshima, Japan. (The technology should have been developed to benefit us, not destroy us.)

19. To replace the CP Co-gen (~20 MW) by Noncombustible Renewable Energy with a Heat Rate given in Table **6.1**, estimate the cooling rate requirement at the cooling tower.

REFERENCES

[1] https://www.epa.gov/sites/production/files/2015-08/documents/coalfired.pdf.

[2] https://www.eia.gov/totalenergy/data/monthly/pdf/sec12_7.pdf

[3] https://en.wikipedia.org/wiki/Heat_of_combustion#Lower_heating_value

[4] https://www.engineeringtoolbox.com/fuels-higher-calorific-values-d_169.html

[5] F. Kreith and S. Krumdieck, "Principles of sustainable energy systems". CRC press, 2013.

[6] Y. A. Cengel and M.A. Boles, "Thermodynamics: An engineering approach", The McGraw-Hill Companies, Inc., New York, 2007.

[7] Y. Wang and K. S. Chen, "PEM fuel cells: thermal and water management fundamentals". Momentum Press, 2013.

[8] L. Drbal, K. Westra, and P. Boston, (Eds.). "Power plant engineering". Springer Science & Business Media, 2012.

[9] P. K. Nag, "Power plant engineering". Tata McGraw-Hill Education, 2002.

[10] S. Kakaç, R. K. Shah and W. Aung, "Handbook of single-phase convective heat transfer", 1987.

[11] A. Bejan and A.D. Kraus, (Eds.). "Heat transfer handbook" (Vol. 1). John Wiley & Sons, 2003.

[12] R. Wang, R. Wang, & Y. Wang, "*Ex-situ* measurement of thermal conductivity and swelling of nanostructured fibrous electrodes in electrochemical energy devices". Thermal Science and Engineering Progress, 100805, 2020.

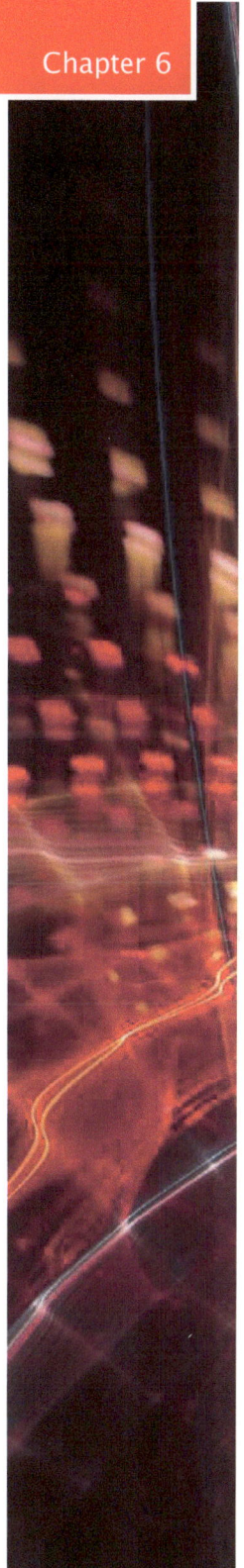

WIND TUNNEL

7.1. INTRODUCTION

Wind tunnels provide controlled air streams to study aerodynamic and fluid flow phenomena. They are frequently used to replicate the actions of an object flying or moving in an air stream, such as airplanes or vehicles. Table **7.1** lists several types of wind tunnels. Most wind tunnels are designed to have a uniform velocity profile in the test section at a low turbulence level. The testing object is placed in the test section for flow visualization and quantitative measurements such as lift and drag forces. To create an air flow, electric fans are usually used to blow air in or out of the tunnel. In large wind tunnels for testing real-size airplanes, rockets, or cars, powerful fans are required to generate a wind speed comparable with real conditions.

A pitot tube is a fluid device widely used to measure flow velocity. It is named after Henri Pitot who invented the device to determine the flow rate of the Seine River in Paris in 1732. The basic structure is a small-diameter tube with its open end facing the flow. It is now a common device for measuring the flow speed in wind tunnels and is widely equipped in airplanes for monitoring their speeds, as shown in Fig. (**S7.1**). Note that there were several fatal aircraft accidents in the past due to ice-plugged pitot tubes or static ports. The aircraft autopilots and real pilots could not cope with the bizarre airspeed or altitude data from plugged ports, causing aircraft uncontrollable dives and stalls.

Table 7.1. Types of wind tunnels.

Type	Remark	Image
Subsonic Tunnel	For a Mach number (M)<0.4. It may be open-return or closed-return type with air moved by axial fans.	
Transonic Tunnel	For 0.75 < M < 1.2. It is designed similarly to subsonic wind tunnels. Perforated/slotted walls are required to reduce wall shock reflection.	
Supersonic Tunnel	For 1.2<M<5. The Mach number and flow are determined by the nozzle geometry. It is usually equipped with a drying or pre-heating facility.	
Hypersonic Tunnel	For 5<M<15. It is designed to simulate the typical features of hypersonic flows, *e.g.* compression shocks, boundary layer, entropy layer, and viscous interaction zones.	

7.2. BERNOULLI EQUATION

Bernoulli's equation is frequently used to solve a set of fluid problems without dealing with the partial differential form of the Navier-Stokes equations. It states that an increase in the fluid velocity occurs simultaneously with a decrease in fluid's static pressure or potential energy. The equation is named after Daniel Bernoulli, who was a Swiss mathematician and physicist and is particularly remembered for his applications of mathematics to fluid mechanics and his pioneering work in probability and statistics [1]. In 1752, Leonhard Euler derived Bernoulli's equation in its usual form [2, 3].

Bernoulli's equation can be derived directly from the Navier-Stokes equation under assumptions. It has various forms determined by the assumptions according to specific problems. Popular assumptions are:

1. Inviscid flow;
2. Steady-state flow;
3. Incompressible flow;
4. Along a streamline or in irrotational flow.

With these assumptions, Bernoulli's equation can be written in concise expression as below:

$$p + \frac{1}{2}\rho u^2 + gz = const \qquad \qquad 1$$

where p is pressure, ρ is density, u is velocity, g is specific weight, and z is elevation. When the streamline is horizontal, *i.e.* z=constant, the above equation changes to:

$$p + \frac{1}{2}\rho u^2 = const \qquad \qquad 2$$

where on the left side, the first term p is referred to as the static pressure, and the second is the dynamic pressure, which is directly related to the fluid velocity u. Their summation denotes the total or stagnation pressure. Fig. (**7.1**) elucidates the static and stagnation pressures at different locations. In practice, we can assume the static pressure at the wall tap in the figure is approximately the same as that near the stagnation point. Thus, their difference, in terms of the hydraulic height ΔH, directly measures the dynamic pressure.

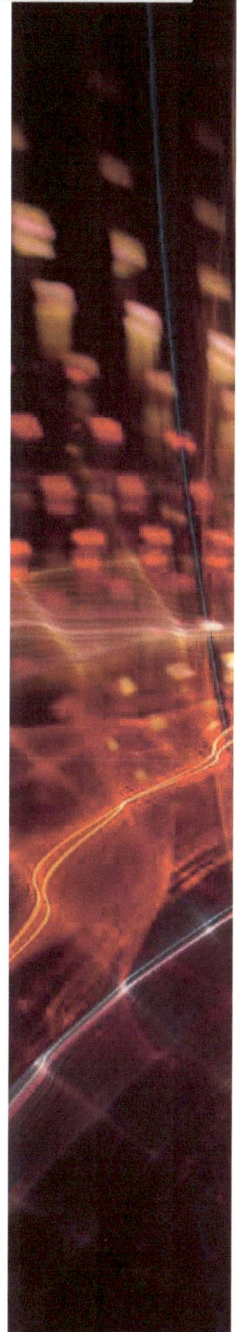

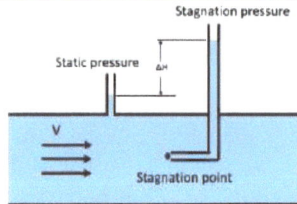

Fig. (7.1). Arrangement of the pitot and static tubes in fluid velocity measurement.

Due to its simplicity without any differential terms, Bernoulli's equation, which describes the basic pressure and velocity relationship, is frequently adopted to study fluid problems and to develop/design flow devices. Successful examples include the pitot tubes that are widely used in wind tunnels and airplanes and the Bernoulli-type flow meters, such as Venturi tubes and Orifice plates, popularly used in industry. The former will be introduced in the next section, while the latter are discussed in Chapter 5.

7.3. PITOT-STATIC TUBE AND VELOCITY MEASUREMENT

A pitot-static tube design is shown in Fig. (**7.2**), in which the pitot tube is in the middle to measure the total pressure, and the static tube is integrated to measure the static pressure. Though the two pressures in the figure are measured at two locations, one can assume the static or total pressures are approximately the same at the two locations. Thus, their difference, as measured directly by a pressure transducer or U-tube manometer in applications, will give the dynamic pressure.

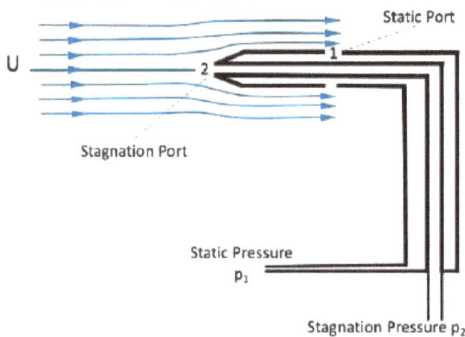

Fig. (7.2). A pitot-static tube to measure U in a uniform flow.

To mathematically derive the fluid velocity in measurement, one can draw a flow streamline that stagnates at the hole in the tip of the pitot tube (labeled 2) and another flow streamline that passes over the hole of the static tube (labeled 1). Note that the upstream is uniform, thus is irrotational flow. Applying Bernoulli's equation will lead to Equation 2 everywhere upstream, including the horizontal streamlines connecting the two points, which gives:

$$p_1 + \frac{1}{2}\rho u_1{}^2 = const = p_2 + 0 \qquad\qquad 3$$

where p_1 and p_2 are the pressures at the static port and the stagnation port, respectively. Again, the total pressure p_2 is also called the stagnation pressure, equal to the sum of the static (p_1) and dynamic ($\frac{1}{2}\rho u_1{}^2$) pressures. Thus, Equation 3 is rearranged to explicitly express u_1 or U:

$$U = u_1 = \sqrt{2\frac{p_2 - p_1}{\rho}} = \sqrt{2\frac{\Delta p}{\rho}} \qquad\qquad 4$$

If the pressure difference Δp is measured in height of liquid in a manometer ΔH or $\Delta p = \rho_l g \Delta H$, the above equation is then changed to the below form:

$$U = K\sqrt{\Delta H} \qquad\qquad 5$$

where K is a constant to express U in the desirable units, such as m/s.

As shown in Equation 5, the velocity calculation from Bernoulli's equation requires the pressure Δp measurement. A large pressure difference is typically measured more accurately than a small one. In measurement, it is often important to assess the sensitivity of the instrument. The sensitivity is the amount of change in the variable one measures per change in the input variable to the instrument. For a pitot tube, the sensitivity can be defined as the change in measured pressure per change in velocity. This is obtained from Bernoulli's equation as:

$$\delta u \sim du = \frac{d\Delta p}{du} = \rho u \qquad\qquad 6$$

Therefore, δu increases with velocity u, *i.e.*, the sensitivity is not constant, which of course, is due to the nonlinear Bernoulli's equation. The above equation shows that pitot tubes are therefore best suited to the measurement of reasonably high flow speeds as on aircrafts.

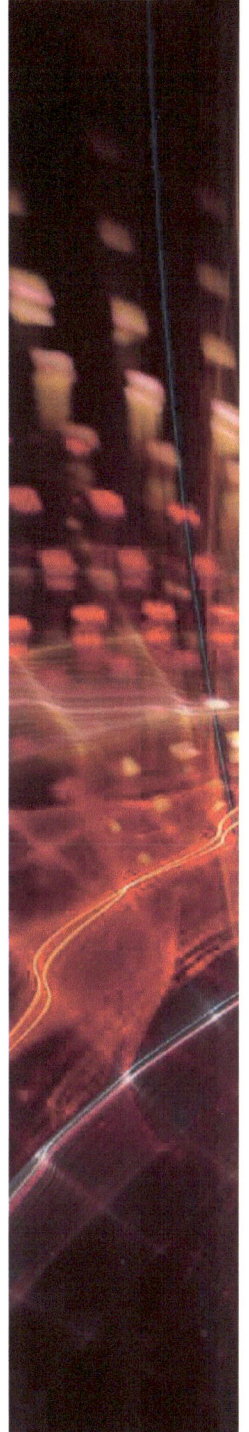

The stagnation pressure is easily measured with a manometer or pressure transducer connected to the center hole of the pitot tube. The static pressure could be measured on the streamline forward of the pitot using a small tube with side holes in it, as shown in Fig. (7.2). However, the insertion of a small tube upstream of the pitot tube will disturb the streamlines. Therefore, the pitot-static tube is developed, which has a concentric tube with static pressure holes in the outer tube at 12 to 24 tube diameters back of the tip. At these locations, the static pressure has "recovered" to the free-stream value from the distortion at the tip. The inner tube is the pitot tube, and both pressures are readily connected to a differential manometer or transducer to obtain Δp.

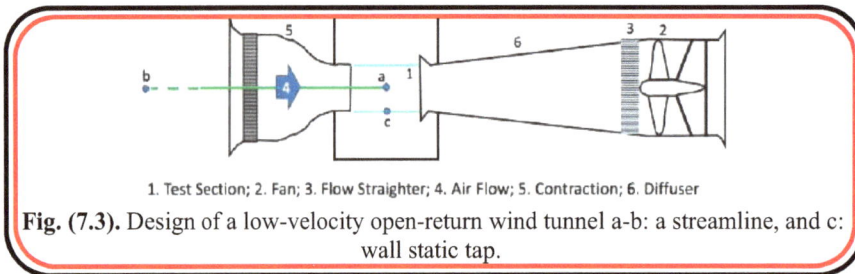

1. Test Section; 2. Fan; 3. Flow Straighter; 4. Air Flow; 5. Contraction; 6. Diffuser

Fig. (7.3). Design of a low-velocity open-return wind tunnel a-b: a streamline, and c: wall static tap.

7.4. WIND TUNNEL BASICS

Fig. (7.3) shows the basic design of a low-speed wind tunnel, which consists of a few major parts, including a contraction section, test section, diffuser, flow straighter, and fan. In some wind tunnels, additional flow straighter is placed before the contraction section to minimize upstream disturbance. In operation, ambient air is drawn into the contraction section and becomes uniform when reaching the test section. The testing object is placed in the test section with proper measurement apparatus. After the test section, air will flow through the diffuser and flow straighter and then out of the wind tunnel. The flow straighter is physically a porous medium consisting of many small identical straight pipes or channels to minimize downstream disturbance by the fan. To apply Bernoulli's equation, a streamline can be virtually drawn horizontally from Point a in the test section to the outside of inlet (labeled Point b), where ambient air is almost at rest, *i.e.* $u_b = 0$:

$$p_a + \frac{1}{2}\rho u_a{}^2 = p_b + \frac{1}{2}\rho u_b{}^2 = p_b \qquad\qquad 7$$

Note that p_b is the ambient pressure, and Point a is an arbitrary point inside the test section outside of the thin boundary layer near the wall. In the test section, when no object is placed, p_a is almost the same, and the flow is almost uniform.

A pitot-static tube is more difficult to make than a simple pitot tube. For wind tunnel application, we recall that for uniform flow, the Bernoulli constant is the same across all the streamlines. If the small pressure drop across the high Reynolds number boundary layer on the walls of the straight test section is ignored, then one can simply put a small hole in the test section's wall to serve as the static pressure tap (Point c as shown in Fig. (**7.3**)), yielding:

$$p_{a,Pitot} = p_c + \frac{1}{2}\rho u_a^2 \qquad\qquad \textbf{8}$$

Note that Points a and c are at different locations with the latter fixed at the wall in the test section. By using this static pressure tap, only a pitot tube is needed for velocity measurement at any point in the test section.

#Example 1: For an air speed of 10 m/s in the test section of Figure 7.3 under standard condition, estimate the total and static pressures at a, and the pressures measured at b and c.

Solution:

$p_{a,Pitot}=1\ atm$

$p_a = p_{a,Pitot} - \frac{1}{2}\rho u_a^2 = 1\ atm - \frac{1}{2}(1.225\ kg/m^3)(10\ m/s)^2$

$=(101,325-61)\ Pa=101,264\ Pa$

$p_b=1\ atm$

From Equation 8:

$p_c = p_{a,Pitot} - \frac{1}{2}\rho u_a^2 = 101,264\ Pa$

The same arrangement is often used on aircrafts, where the pitot tube and static pressure tap are often located at different places, as shown in Fig. (**S7.1**). The static pressure taps are located on the fuselage at locations where the measured pressure is close to the free-stream pressure.

The static ports on both sides of the fuselage are manifolded together so that correct pressure is obtained when the aircraft yaws. The static pressure is also used to calculate the altitude referenced to a standard atmosphere. The pitot tubes stick out into the air stream and are heated in case of icing conditions. The next time you fly on an airliner, look at the pitot tubes at the front of the plane when the plane is parked at the gate. There are several for redundancy. Depending on the aircraft model, you may also see the flush static pressure ports, which are usually in an unpainted circle on the fuselage with a warning label to keep the ports clean.

7.5. WIND TUNNEL EXPERIMENT

As an example to explain wind tunnel testing, a small open-return wind tunnel is introduced in this section, as shown in Fig. (**S7.3** or **7.3**), in which room air is drawn into the test section through a smooth contraction section. The blower is located downstream of a honeycomb flow straightener to reduce swirl. Test section speed is varied by throttling the outlet of the constant-speed blower. A pitot tube on a traverse is connected to a liquid manometer, which contains a special low volatility oil and has a scale calibrated in inches of water. In practice, an electronic transducer is usually used to convert the pressure to a digital signal. The air density (1.225 kg/m^3) will be required in velocity calculation. A sample of testing data is given in Table **7.2**.

Table 7.2. Sample of experimental data.

Free Stream Profile		Wake Flow Profile	
Position (m)	ΔH (m)	Position (m)	ΔH (m)
0.00320	0.03048	0.00320	0.03556
0.00574	0.03556	0.01082	0.04064
0.00828	0.03810	0.01844	0.04191
0.01082	0.03937	0.02606	0.04191
0.01336	0.04039	0.03368	0.04191
0.01590	0.04064	0.04130	0.04064
0.01844	0.04064	0.04384	0.04064
0.02098	0.04064	0.04638	0.04064
0.02352	0.04064	0.04892	0.03810
0.02606	0.04064	0.05146	0.03556
0.02860	0.04064	0.05400	0.02794
0.03368	0.04064	0.05654	0.02032
0.03876	0.04064	0.05908	0.01016
0.04384	0.04064	0.06162	0.00000
0.04892	0.04064	0.06416	0.00000
0.05400	0.04064	0.06670	0.00508
0.06670	0.04064	0.06924	0.01016
0.07940	0.04064	0.07178	0.02286
0.09210	0.04064	0.07432	0.02921
0.10480	0.03937	0.07686	0.03683
0.10734	0.03810	0.07940	0.03810
0.10988	0.03810	0.08194	0.04064
0.11242	0.03810	0.08956	0.04064
0.11496	0.03683	0.09718	0.04064
0.11750	0.03683	0.10480	0.04064
0.12004	0.03302	0.11242	0.04064
0.12258	0.02921	0.12004	0.03937
0.12512	0.02921	0.12512	0.03302

7.5.1. Pressures in Wind Tunnel

To understand how the pitot tube, static tube, and wall static tap work, position the pitot-static tube in the center of the tunnel to:

1. Measure the pressure difference between the pitot pressure from the pitot-static tube and the wall static tap (c). Do not hook up the static tube of the pitot-static tube to the manometer.

2. Measure the pressure difference between the pitot pressure and static tube pressure from the pitot-static tube. Seal the wall static tap before the measurement.

3. Carefully remove the plastic tubing line from the pitot tube. Observe and record the pressure difference between the open tube and the wall static tap (c).

#Example 2: Think about the results from (1) and (3). If the pressures are the same within experimental uncertainty, why?

Solution:

1 measure the pressure difference between the total pressure (1 atm) and ambient pressure (1 atm);

2 measure the pressure difference between the total pressure (1 atm) and static pressure;

3 measure the pressure difference between the ambient pressure (1 atm) and static pressure.

Thus, 2 and 3 are the same, while 1 gives 0 Pa reading.

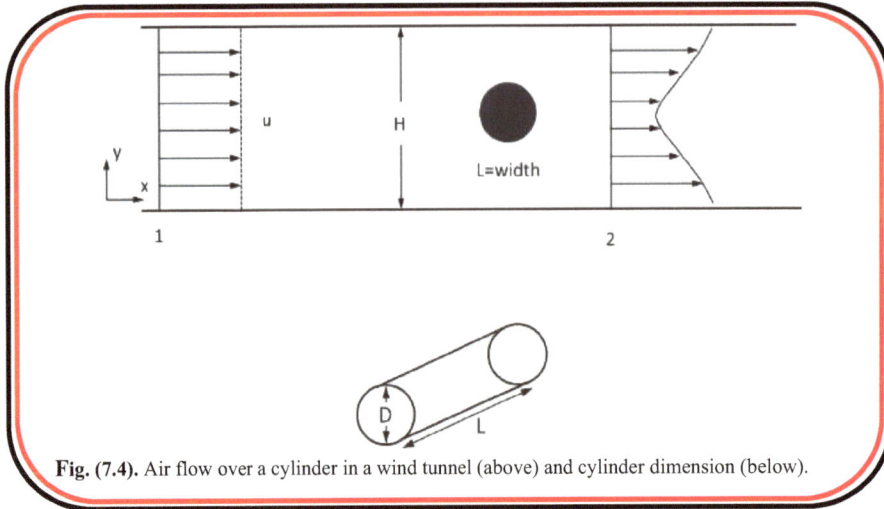

Fig. (7.4). Air flow over a cylinder in a wind tunnel (above) and cylinder dimension (below).

7.5.2. Velocity Profile

In the wind tunnel, the test section has a uniform velocity when no testing object is placed in. To verify this, we will measure local air velocity. The pitot tube-static wall tap arrangement will be used to traverse the test section and obtain the velocity profiles across the section. Non-uniform spacing of the profile data points with higher resolution may be taken to see the sharp velocity change in the boundary layers. In the experiment, i) mount the pitot tube on the height gauge and note the scale readings and units. ii) set a reasonable tunnel airspeed (for example, the manometer at least 1.0 inch water). iii) conduct measurement of velocity at each position. The measured data are in the form of (y_i, u_i), where y_i is the physical location of the pitot tube above the bottom wall of the test section with $i=0$ and N representing the lowest ($y = -\frac{H}{2}$) and highest ($y = \frac{H}{2}$) locations in the test section. One can then plot the velocity profile using the data, which should show a flat profile in the mainstream and a sharp change in the boundary layer, as shown in Figure S7.4. To use the data for calculating the flow rate $\dot{V}$ and the average velocity, one can integrate the velocity over y:

$$\dot{V} = \int_{-\frac{H}{2}}^{\frac{H}{2}} L \cdot u(y)dy = L \cdot \sum_N \bar{u}_{i+\frac{1}{2}}(y_{i+1} - y_i) = L \cdot \sum_N \frac{u_{i+1}+u_i}{2}(y_{i+1} - y_i) \quad \longrightarrow \quad 9$$

where L is the test section width, as shown in Fig. (**7.4**). In the following derivation, we label the velocity profile 1 or free stream, as measured at the location 1 in Fig. (**7.4**). Assume $u(y)$ is uniform, *i.e.* ignoring the boundary layers:

$$\dot{V} \approx L \cdot u_1 \cdot H = L \cdot U \cdot H \qquad \longrightarrow \qquad 10$$

where U is the free stream velocity, similar to that of flying aircrafts. Fig. (**S7.6**) gives the sample of Matlab codes for flow rate and average velocity calculation. The mass flow rate can then be obtained by:

$$\dot{m} = \dot{V} \cdot \rho \qquad \longrightarrow \qquad 11$$

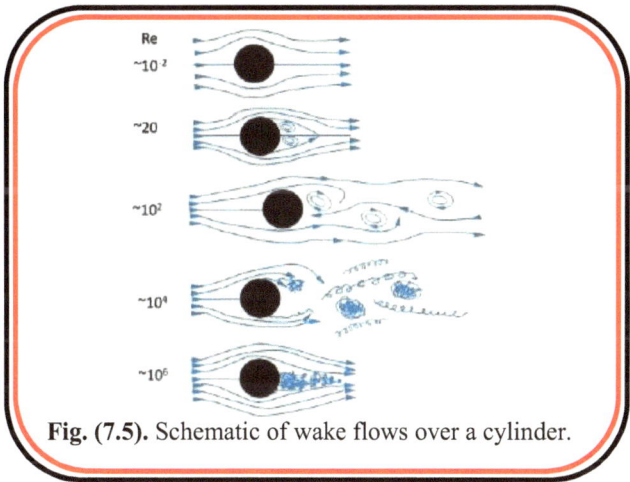

Fig. (7.5). Schematic of wake flows over a cylinder.

7.5.3. Cylinder Wake

Wake flow is the disturbed flow downstream of a moving solid body in a fluid, as shown in Figs. (**S7.2 and 7.5**). As a basic example of measuring air wake flow, a small cylinder is tested: i) horizontally insert a cylinder in the middle of the test section upstream of the pitot tube. Prior to the testing, one should make a preliminary traverse of the cylinder wake to ensure that the minimum pitot pressure is measurably above 0. If the pitot is too close to the cylinder, local reverse flow may be present. A pitot tube does not work in such a situation.

Think about why? If the minimum pitot pressure is too low, move the cylinder to another location farther upstream. ii) conduct measurement of velocity at each position. To distinguish from the free stream profile, this velocity profile is labeled 2, as if measured at location 2 in Fig. (**7.4**).

Carefully profile the wake of the cylinder by recording the location of the pitot tube from the position gauge y and the manometer pressure ΔH. Obtain closely-spaced readings where the velocity gradients are large. One will then obtain a set of data, labeled by (y_{2i}, u_{2i}) to distinguish the free stream profile data (y_{1i}, u_{1i}) at location 1. For the control volume 1-2 as shown in Fig. (**7.4**), the mass balance will give:

$$L \cdot U \cdot H = \int_{A_1} u_1 dA_1 = \int_{A_2} u_2 dA_2 = L \int_{-\frac{H}{2}}^{\frac{H}{2}} u_2 dy \qquad \Longrightarrow \qquad 12$$

To obtain a comprehensive picture of the wake flow, one can repeat the measurement at another distance from the cylinder, *e.g.* location 3 for a data set (y_{3i}, u_{3i}), which also satisfies the above mass balance.

7.5.4. Drag Coefficient (C_D)

The drag force that air flow imposes on an airplane is important to airplane design optimization and fuel consumption reduction. In this subsection, the measurement data of the cylinder wake flow will be used as an example to show how to calculate the drag force and drag coefficient (C_D).

In Fig. (**7.4**), the momentum balance of the control volume in the test section between 1 and 2 gives:

$$\sum F_X = \int_{①} \rho u_1 u_1 dA_1 - \int_{②} \rho u_2 u_2 dA_2 \qquad \Longrightarrow \qquad 13$$

Assume $u_1 = U$, the mass balance, Equation 12, gives:

$$\int_{①} \rho u_1 u_1 dA_1 = \rho U \int_{①} u_1 dA_1 = \rho U \int_{②} u_2 dA_2 \qquad \Longrightarrow \qquad 14$$

One will then reach:

$$\sum F_X = \rho U L \int_{\frac{H}{2}}^{\frac{H}{2}} u_2 dy - \rho L \int_{\frac{H}{2}}^{\frac{H}{2}} u_2^2 dy = \rho L \int_{\frac{H}{2}}^{\frac{H}{2}} (U u_2 - u_2^2) dy =$$

$$\rho L \int_{\frac{H}{2}}^{\frac{H}{2}} u_2 (U - u_2) dy = \rho L U^2 \int_{\frac{H}{2}}^{\frac{H}{2}} \frac{u_2}{U} \left(1 - \frac{u_2}{U}\right) dy \quad \longrightarrow \quad \textbf{15}$$

where the integrand $\frac{u_2}{U}\left(1 - \frac{u_2}{U}\right)$ is called the momentum deficit. Fig. (**S7.5**) plots the profiles of the wake flow and momentum deficit. We now use $\frac{u}{U}\left(1 - \frac{u}{U}\right)$ to replace the integrand as it is for general wake flows. To calculate the dimensionless force and pressure, we can define C_D as:

$$C_D = \frac{\sum F_X / A_c}{1/2\rho U^2} \quad \longrightarrow \quad \textbf{16}$$

where A_c is the project area of the cylinder facing the air flow, which equal to D multiplied by L, as shown in Fig. (**7.4**). Then,

$$C_D = \frac{\rho L U^2 \int_{-\frac{H}{2}}^{\frac{H}{2}} \frac{u}{U}\left(1 - \frac{u}{U}\right) dy}{DL(1/2\rho U^2)} = \frac{2}{D} \int_{-\frac{H}{2}}^{\frac{H}{2}} \frac{u}{U}\left(1 - \frac{u}{U}\right) dy \quad \longrightarrow \quad \textbf{17}$$

Define a dimensionless height $\eta = \frac{y}{D}$ and rewrite Equation 17 as:

$$C_D = 2 \int_{-\frac{H}{2D}}^{\frac{H}{2D}} \frac{u}{U}\left(1 - \frac{u}{U}\right) d\eta \quad \longrightarrow \quad \textbf{18}$$

In the experiment, a small cylinder is used, *i.e.* H>>D, thus we can assume $\frac{H}{2D} \to \infty$, yielding:

$$C_D = 2 \int_{-\infty}^{\infty} \frac{u}{U}\left(1 - \frac{u}{U}\right) d\eta \quad \longrightarrow \quad \textbf{19}$$

Note that $\frac{H}{2D} \to \infty$ makes the wind tunnel testing equivalent to the real condition of a flying object in ambient air. Also, C_D is independent of any tunnel physical dimension. In general wind tunnel testing, it is required that the tunnel size is much larger than the testing object, *i.e.* H>>D, in order to obtain data comparable to practical conditions. Fig. (**S7.7**) gives a sample of Matlab codes to calculate C_D.

#Example 3: assuming a cylinder horizontally falls has a C_D of 1.2 with 30 cm in diameter (D) and 170 cm in length (L). Estimate the drag force at a falling speed of 100 m/s.

Solution:

$A_c = DL = 0.51\ m^2$

Drag force: $\sum F_X = \frac{1}{2}\rho U^2 C_D A_c = 0.5(1.225\ kg/m^3)(100\ m/s)^2 1.2(0.51\ m^2) = 3,750\ kg\ m/s^2$
$= 3,750\ N$

The dimension of the cylinder is comparable to a human body. It is seen that a sky-diving person cannot reach 100 m/s without a parachute because the drag force is far over his/her weight at speed.

Furthermore, the wake flow over a cylinder is a function of Re, as shown in Fig. (**7.5**), which displays typical wake flow patterns. Each pattern should lead to a specific trend of C_D change. Fig. (**7.6**) shows C_D of a sphere and cylinder as a function of Re. Additionally, the wake flow is assumed two dimensional (2D) for a cylinder of a large ratio of its length to diameter. As the cylinder becomes short, the 2D assumption fails, which reduces the value of C_D, as shown in Table **7.3**.

Fig. (7.6). Drag coefficients of spheres and cylinders [4].

Table 7.3 Drag coefficients of a cylinder with a length L and diameter b_0 at Re=88,000 [5,6].

$\left(\dfrac{L/b_0}{\text{Infinity}}\right)$	$\left(\dfrac{C_D}{1.2}\right)$
40	0.98
20	0.92
10	0.82
5	0.74
3	0.74
2	0.68
1	0.63

For further readings regarding wind tunnels, air flow, drag forces, and experiments, we refer the interested readers to the references in [5-9].

7.6. QUESTIONS

1. Using Bernoulli's equation. Are the pressures obtained from the Pitot-wall tap and open tube-wall tap measurements the same? Explain in terms of Bernoulli's equation. A sketch may help.

2. Boundary Layer. Why and under what conditions may we confidently neglect the pressure drop across the thickness of the boundary layer on the wind tunnel wall when using the wall static pressure tap?

3. Velocity Profile. Plot the velocity profile in dimensional form and non-dimensional form with the velocity normalized with the center-line, maximum value, and the test section position by the section width. Integrate the profile to obtain the flow rate and average velocity and plot the average velocity as a separate line on the appropriate plot.

4. Wake and Drag. Plot the velocity distribution in the cylinder wake with the velocity normalized with the free-stream value and the distance by the diameter of the cylinder.

5. Plot the normalized momentum deficit *versus* non-dimensional distance ($\eta = \frac{y}{D}$). Obtain the drag on the cylinder from the momentum deficit profile by numerical integration. Calculate the drag coefficient and plot it in Fig. (**7.6**).

6. In the experiment, if the pitot is too close to the cylinder, the local reverse flow may be present. A pitot tube does not work in such a situation. Why?

7. $\frac{H}{2D} \to \infty$ makes the wind tunnel testing equivalent to real condition of airplane flying in ambient air. Why?

8. In Equation 5, please express K for U in a unit of: m/s, in per hour, and mm/min for ΔH in [mm].

9. Describe the wake flow of a spherical ball.

10. Describe how the measurement strategy and equations will change if the measurement is for a spherical ball instead of a cylinder.

11. Set up 2D CFD simulation of the experiment and plot the flow fields (or velocity profiles) before and after the cylinder.

12. Describe how the length of the cylinder impacts the drag force and C_D.

13. Why in the C_D definition the projected area of the cylinder is used, instead of the surface area?

14. Describe possible errors or uncertainties associated with the control volume analysis using Fig. (**7.4**).

16. In Fig. (**7.6**), C_D decreases with an increasing Re at the beginning; does that mean the drag force decrease as well?

17. Estimate the equivalent speed of a falling cylinder in Example 3.

18. Estimate the equivalent speed of a falling rain droplet of a diameter 1 mm if $C_D=10$ or 60.

REFERENCES

[1] D. Bernoulli, Hydrodynamica. "Britannica online encyclopedia", ed: Retrieved 2008-10-30, 2008.

[2] S. R. On, "The history of fluid dynamics", The Handbook of Fluid Dynamics, 1998.

[3] O. Darrigol and U. Frisch, "From Newton's mechanics to Euler's equations", Physica D: Nonlinear Phenomena, vol. 237, no. 14-17, pp. 1855-1869, 2008.

[4] H. Schlichting, "Boundary layer theory", McGraw Hill, 1968.

[5] C. F. Heddleson, D. L. Brown and R. T. Cliffe, "Summary of drag coefficients of various shaped cylinders". General Electric Co Cincinnati Oh, 1957.

[6] J.Knuddsen and D. L. Katz, "Fluid dynamics and heat transfer", 1958.

[7] Y. Wang and K. S. Chen, "PEM fuel cells: thermal and water management fundamentals". Momentum Press, 2013.

[8] S. C. Cho, Y. Wang, and K. S. Chen, "Droplet dynamics in a polymer electrolyte fuel cell gas flow channel: Forces, deformation, and detachment. I: Theoretical and numerical analyses", Journal of Power Sources, vol. 206, pp. 119-128, 2012.

[9]   S. C. Cho, Y. Wang, and K. S. Chen, "Droplet dynamics in a polymer electrolyte fuel cell gas flow channel: Forces, deformation, and detachment. II: Comparisons of analytical solution with numerical and experimental   results. Journal of Power Sources" , Journal of Power Sources, vol. 210, pp.191-197, 2012.

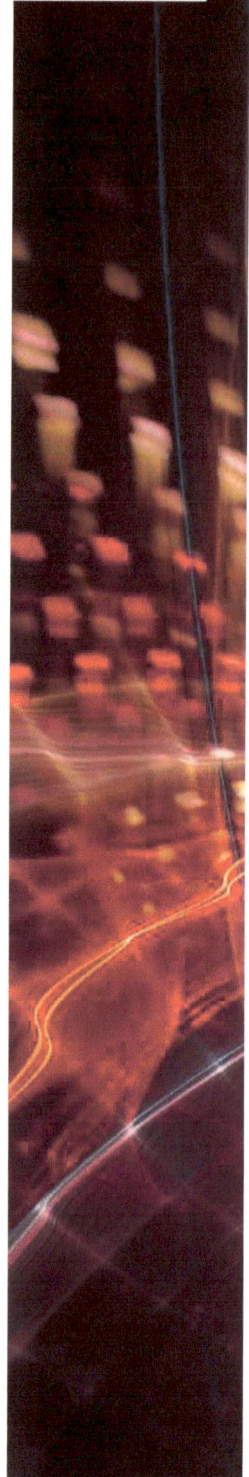

OTTO AND DIESEL CYCLES

8.1. INTRODUCTION

Otto and Diesel cycle engines play an important role in our transportation and energy use. They are typically reciprocating heat engines that convert the thermal energy from fuel combustion to mechanical energy in the form of piston movement. The mechanical energy further drives a vehicle over a distance. The Otto and diesel cycle engines are the most common engine in passenger cars, light trucks, and other applications where small (10 Hp) to medium power (500 Hp) is required. Some large turbo supercharged radial aircraft engines reach 5,000 Hp. Applications of small power, such as lawnmowers and hand-held devices like trimmers and chain saws, require a level of 100-1,000 W power. Typical values of their thermal efficiency are 30-35% for Otto cycle engines and 30-40% for Diesel engines. Small utility-type engines may have ~20% efficiency due to simple design and control. While the basic principles of these reciprocating engines have not changed significantly since invention, advances in fuel induction, ignition systems, and exhaust emission controls have improved economy and performance and reduced pollution.

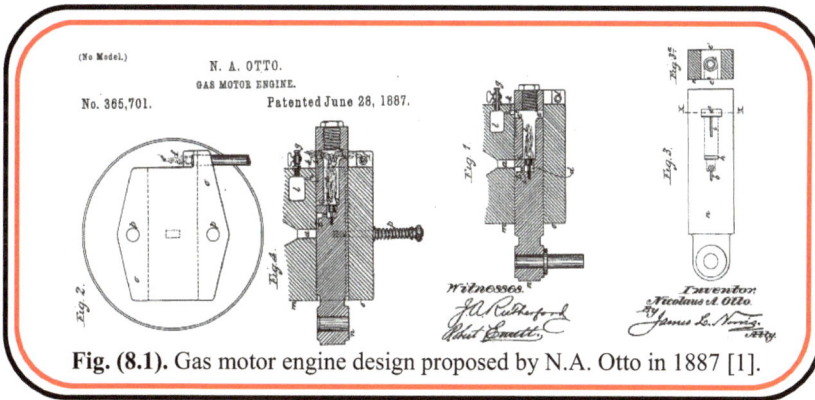

Fig. (8.1). Gas motor engine design proposed by N.A. Otto in 1887 [1].

8.2. OTTO AND DIESEL CYCLES

A working four-stroke engine with an igniting apparatus was first developed by a German engineer Nicolaus Otto. Fig. (**8.1**) shows Otto's original patent. Gasoline engines, commonly used in passenger cars, are often called Otto cycle engines, which ignite the compressed mixture of gasoline vapor and air using spark plugs. Different from gasoline engines, diesel engines inject fuel into the compressed air in the cylinder, which is then ignited by the hot air. The ideal closed-system Otto cycle consists of four basic processes that occur in two strokes of the piston in the cylinder. It differs from the Carnot cycle in the processes of heat interactions, which occur under constant volume instead of constant temperature. For the diesel cycle, the heat addition is under constant pressure due to the relatively long combustion process comparing with the Otto cycle. Table **8.1** summarizes the ideal Otto and diesel cycle processes and thermodynamic equations.

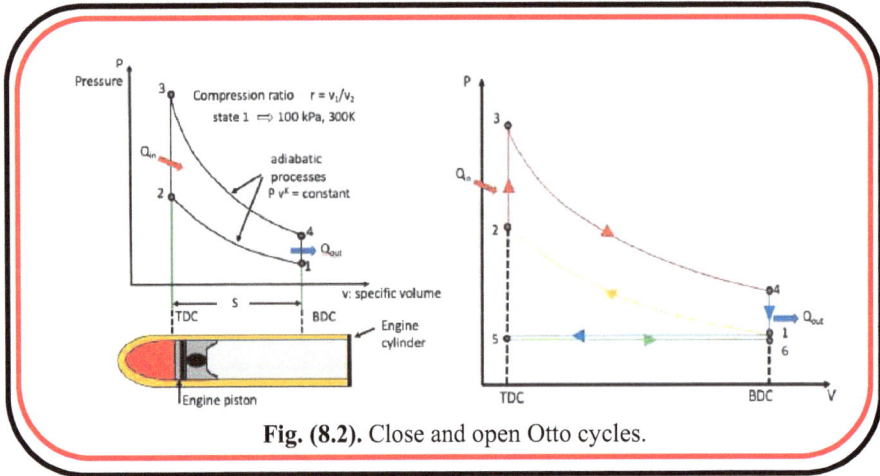

Fig. (8.2). Close and open Otto cycles.

The real Otto cycle is an open system with two extra strokes required in the practical implementation of the Otto cycle - one extra stroke at the end of the expansion (power) stroke to exhaust the burnt fuel followed by an extra stroke to intake a new fuel/air mixture, as shown in Fig. **(8.2)**. The theoretical thermal efficiency η_{th} is given by:

$$\eta_{th} = \frac{W_{out}}{Q_{in}} = 1 - \frac{Q_{out}}{Q_{in}} \qquad\qquad 1$$

Table 8.1. Otto and diesel cycle processes and equations.

Otto Cycle	**Energy Eq. (1st law)**	**Entropy Eq.**	**Ideal Path**
Compression	$u_2 - u_1 = -_1w_2$	$s_2 - s_1 = (0/T) + 0$	$q = 0, s_1 = s_2$
Combustion	$u_3 - u_2 = q_H$	$s_3 - s_2 = \int dq_H/T + 0$	$v_3 = v_2 = C$
Expansion	$u_4 - u_3 = -_3w_4$	$s_4 - s_3 = (0/T) + 0$	$q = 0, s_3 = s_4$
Heat rejection	$u_1 - u_4 = -q_L$	$s_1 - s_4 = -\int dq_L/T + 0$	$v_4 = v_1 = C$

(Table 8.1) cont.....

Diesel Cycle	Energy Eq. (1st law)	Entropy Eq.	Ideal Path
Compression	$u_2 - u_1 = -_1w_2$	$s_2 - s_1 = (0/T) + 0$	$q = 0, s_1 = s_2$
Combustion	$u_3 - u_2 = q_H - _2w_3$	$s_3 - s_2 = \int dq_H/T + 0$	$P_3 = P_2 = C$
Expansion	$u_4 - u_3 = -_3w_4$	$s_4 - s_3 = (0/T) + 0$	$q = 0, s_3 = s_4$
Heat rejection	$u_1 - u_4 = -q_L$	$s_1 - s_4 = -\int dq_L/T + 0$	$v_4 = v_1 = C$

8.2.1. Otto Cycle Engine

Through the isentropic compression-expansion relationships for a calorically perfect gas (CPG), the Otto cycle efficiency $\eta_{th,\text{Otto}}$ can be derived as below:

$$\eta_{th,\text{Otto}} = 1 - \frac{1}{r_v^{k-1}} \qquad\qquad\qquad\qquad\qquad 2$$

where k is the ratio of specific heats and r_v is the volumetric compression ratio.

#Example 1: for k = 1.4 and = 8.5, calculate the Otto cycle efficiency! What should r_v be in order to achieve $\eta_{th,\text{Otto}}$=90%?

Solution:

Using Equation 2,

$$\eta_{th,\text{Otto}} = 1 - \frac{1}{8.5^{(1.4-1)}} = 57.5\%$$

In order to achieve 90% efficiency,

$$1 - \frac{1}{r_v^{k-1}} = 90\% \rightarrow r_v = 300$$

$\eta_{th,\text{Otto}}$ is the maximum possible efficiency under this engine configuration, which is less than the Carnot efficiency operating between the maximum and minimum temperatures. It is against this maximum that actual engines should be compared. Typical peak efficiencies of modern passenger car engines are approximately 30%. Under off-peak conditions (such as maneuvering in a parking lot) it drops to 10-15%.

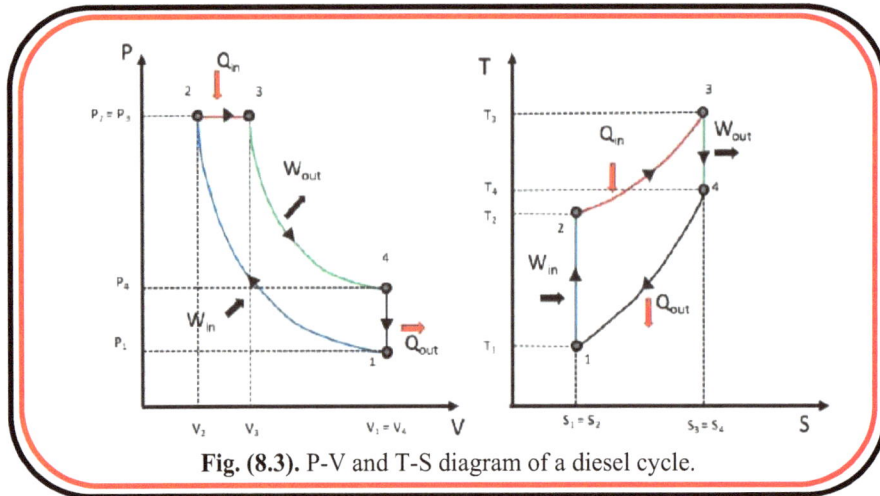

Fig. (8.3). P-V and T-S diagram of a diesel cycle.

8.2.2. Diesel Cycle Engine

In the Otto cycle, the air-fuel mixture is compressed by the piston, which increases its temperature to a level lower than the fuel's auto-ignition temperature. Combustion is then ignited by a spark plug when reaching the top dead center (TDC) of the piston movement. Fig. (**8.3**) shows the P-V and T-S diagrams. In diesel engines, only air is compressed in the cylinder, which increases the air temperature above the fuel's auto-ignition temperature at the TDC so that the injected fuel can be ignited by the hot air.

Through the isentropic compression-expansion relationships for a calorically perfect gas (CPG) and cold-air standard assumptions, the diesel cycle efficiency $\eta_{th,\text{Diesel}}$ can be derived as:

$$\eta_{th,Diesel} = 1 - \frac{1}{r_v^{k-1}}\left[\frac{r_c^k - 1}{k(r_c - 1)}\right] \qquad \textbf{3}$$

where r_c is the cutoff ratio, defined as the ratio of the cylinder volumes after and before the combustion process [2]. Note that typically the quantity in the bracket is greater than 1 thus $\eta_{th,Diesel} < \eta_{th,Otto}$ under the same compression ratio. However, diesel engines usually operate at much higher compression ratios and thus are more efficient than the spark-ignition Otto cycle engines at full loads. In addition, due to significant throttling losses at part load, Otto cycle engines suffer a much greater loss of efficiency at part-load conditions than comparable diesel engines.

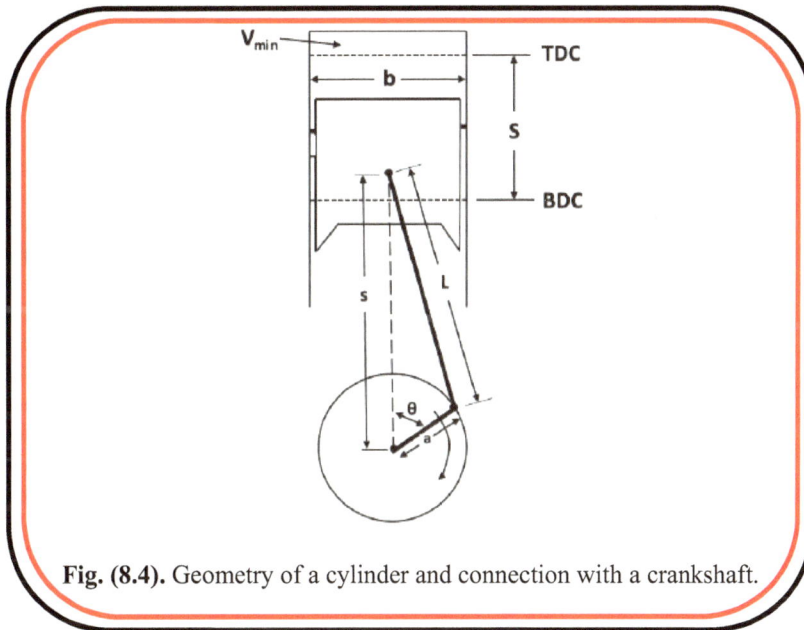

Fig. (8.4). Geometry of a cylinder and connection with a crankshaft.

8.2.3. Engine Basics

Implementation of the open system Otto cycle is shown in Fig. (**S8.1**), where the up-and-down motion of the piston in the cylinder is transformed into rotary motion through the connecting rod to the crankshaft. Figure 8.4 shows the geometry of a cylinder and its connection with a crankshaft. The volume in the piston-cylinder is a function of the angular position of the crankshaft. Top dead center (TDC) refers to the crankshaft angle $\theta = 0°$ where the piston is at the top of its stroke (of the # 1 cylinder in a multi-cylinder engine). The volume at this position is minimum, often called the clearance volume V_{min}. The bottom dead center (BDC) refers to the crankshaft position at $\theta = 180°$. The volume is maximum V_{max} at this condition. Fig. (**8.2**) draws the piston locations in the P-V diagram. The compression ratio r_v is defined as the ratio of the maximum to minimum volumes:

$$r_v = \frac{V_{max}}{V_{min}} \qquad\qquad\qquad\longrightarrow \qquad 4$$

The displacement volume V_d is the difference between the maximum and minimum volumes:

$$V_d = V_{max} - V_{min} = \frac{\pi}{4} b^2 S \qquad\longrightarrow \qquad 5$$

where b is the cylinder bore diameter and S is the piston stroke. Because a reciprocating engine uses a slider-crank mechanism, the cylinder volume is not a simple function of the crankshaft angle. For typical geometries where the connecting rod length L is large compared to the stroke (ε = S/L << 1), the displacement function is approximate:

$$\tilde{V} = 1 + \frac{r_v - 1}{2}(1 - \cos\theta) \qquad\longrightarrow \qquad 6$$

where $\tilde{V} = \frac{V(\theta)}{V_{min}}$. The engine speed R_s refers to the rotational speed of the crankshaft, usually revolutions per minute (RPM). The mean piston speed is then:

$$\tilde{U}_p = 2SR_s \qquad\qquad\longrightarrow \qquad 7$$

The indicated work W_i is the net work produced in the cylinder by gas acting against the piston during the compression and expansion strokes. The part of this work that is available at the crankshaft is the brake work W_b. $W_b < W_i$ and the difference is due to the frictional work W_f and pumping work during intake and exhaust strokes W_p:

$$W_b = W_i - W_f - W_p \qquad\longrightarrow \qquad 8$$

In vehicles, accessories such as the water pump, electrical generator, air-conditioning compressor, *etc.*, are external to the engine, and part of W_b is used for these loads. The actual work or power available at the rear wheels is measured with a chassis dynamometer.

For a single-cylinder engine, the indicated work is given by:

$$W_i = \oint P \, dV \qquad \qquad 9$$

where P is the cylinder pressure and V is the cylinder volume. A continuous plot of P in the cylinder and V as a function of time throughout the cycle is called an Indicator Diagram and W_i can be obtained from graphical integration of the area enclosed by the P-V trace. Note V can be obtained from the angle of the crankshaft as given in Equation 6. Indicator diagrams are used on large, slow-turning steam piston engines to measure the work while running. An indicator diagram example from a gasoline engine is shown in Fig. (**S8.3**), where a fast-response pressure transducer is installed in the cylinder and V is obtained from a pick-off of the crankshaft angle, θ. Both the power-producing part of the cycle and the pumping part are clearly shown on the indicator diagram. The directions of the paths around the cycle have to be kept in mind when calculating the total indicated work: the top portion is clockwise, representing work done, while the bottom portion (exhaust pumping against atmospheric pressure and intake vacuum relative to atmospheric pressure) is counter-clockwise, representing work. The total indicated work is the algebraic sum of the two work interactions.

Usually, engine behavior is described in terms of power and torque rather than work. The torque τ is simply work per unit crank rotation (in radians). For a four-stroke engine, the indicated torque and power are given by:

$$\tau_i = W_i/4\pi \text{ and } \dot{W}_i = 2\pi\tau_i R_s \qquad \qquad 10$$

Brake power and friction power are similarly defined. To normalize out the engine size, power is often characterized by the mean effective pressure. The mean effective pressure is the work done per unit displacement volume. For a four-stroke engine, the brake means effective pressure (bmep) is given by:

$$bmep = \frac{4\pi\tau_b}{V_d} = \frac{2\dot{W}_b}{V_d R_s} \qquad \qquad 11$$

The units of bmep are, of course, those of pressure, *e.g.*, psi in the US. Note that bmep is the constant pressure acting through the displacement volume that would produce the same work, the area on the P-V indicator diagram, as occurs in the actual cycle between the volume limits V_{min} and V_{max}.

The brake-specific fuel consumption (bsfc) is another measure of engine efficiency. It is defined as the fuel mass flow rate $\dot{m}_f$ divided by the power output:

$$bsfc = \frac{\dot{m}_f}{\dot{W}_b} \qquad\qquad\qquad\qquad 12$$

It is dimensional, and the US unit is pounds/Hp-hr. Typical values are about 0.35–0.5 pounds/Hp-hr.

#Example 2: Calculate bmep and bsfc for the first data in case#1 in Figure 8.7, along with the brake mean effective torque. The displacement volume and gasoline density are 19.44 in³ and 720kg/m³.

Solution:

For the first data in the case#1,

$$\dot{W}_b = 2\ Hp = 1.49\ kW\ and\ R_s = 930\ RPM = 15.5\ /s$$

$$V_d = 19.44\ in^3 = 0.0003186\ m^3$$

$$bmep = \frac{2\dot{W}_b}{V_d R_s} = 2\frac{1490\ W}{0.0003186\ m^3 * 15.5\ /s} = 603,400\ Pa\ or\ 0.6034\ MPa.$$

From Figure 8.7, the readings for consumed fuel and duration are 50 mL and 159.2 s (or 0.044 hr); thus, the fuel volumetric consumption rate is given by:

$$50\ mL/0.044\ hr = 1.1\ L/hr$$

The fuel mass consumption rate is given by:

$$\dot{m}_f = 1.1L/hr \cdot (0.001\ m^3/L) \cdot 720kg/m^3 = 0.792\ kg/hr = 1.75\ lbs/h.$$

Then,

$$bsfc = \frac{\dot{m}_f}{\dot{W}_b} = \frac{1.75 \ lbs/hr}{2 \ Hp} = 0.875 \ pounds/Hp\text{-}hr.$$

From Equation 11, the brake mean effective torque is given by:

$$\tau_b = \frac{bmep}{4\pi} V_d = 15.3 \ N \cdot m.$$

8.3. ENGINE ANALYSIS

8.3.1. Heat Balance and Fuel Efficiency

Heat balance can be set up for the open-system analysis of engines, as shown in Fig. (**8.5**). It goes further than the determination of the thermal efficiency, which essentially accounts for only one item in the heat balance. The items in the heat balance include:

1. brake work
2. the heat of the fuel expressed as the LHV (lower heating value)
3. sensible heat in the dry exhaust gases
4. heat losses (convective and radiative heat transfer, *etc.*) to the ambient environment.

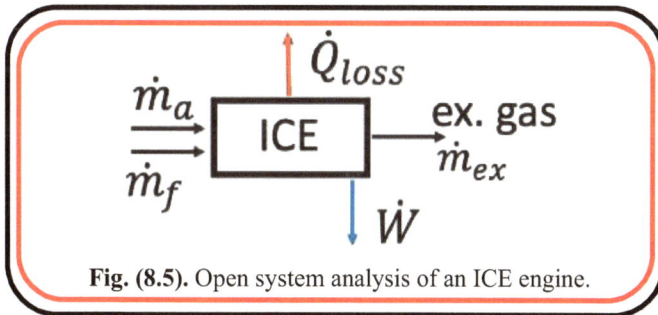

Fig. (8.5). Open system analysis of an ICE engine.

In Fig. (**8.5**), two equations can be set up from the mass balance and first law, as below:

Mass balance: $\dot{m}_a + \dot{m}_f = \dot{m}_{ex}$ ➡️ 13

First law: $\dot{m}_a h_a + \dot{m}_f h_f = \dot{m}_{ex} h_{ex} + \dot{W} + \dot{Q}_{loss}$ ⟶ **14**

To obtain a heat balance on the engine, we can measure the temperature of the exhaust using a thermocouple. The inlet air is assumed to be at room temperature. The fuel injection rate timing the heating value of the fuel will give the heat addition rate. The LHV is frequently used because the water product in the exhaust usually exists in the vapor phase, and we can express LHV approximately by:

$LHV \cdot \dot{m}_f = \dot{m}_a h_a + \dot{m}_f h_f - \dot{m}_{ex} h_{ex,\ 77°F}$ ⟶ **15**

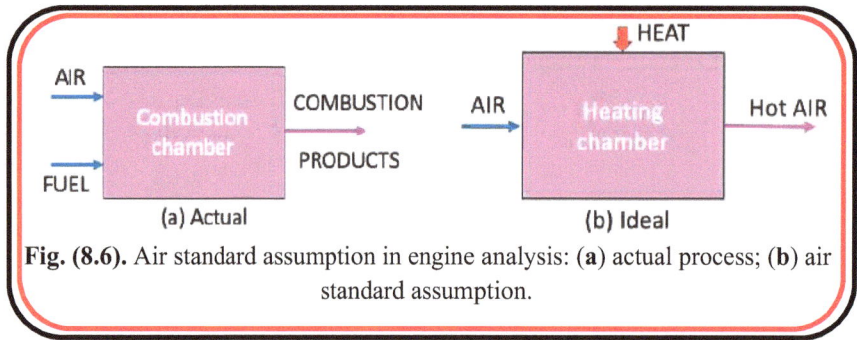

Fig. (8.6). Air standard assumption in engine analysis: (**a**) actual process; (**b**) air standard assumption.

Engine combustion is a complex process involving many elementary reactions and species transports. To simply the system and analysis, the air standard assumption frequently applies, described by:

1.) The working fluid is air, which continuously circulates in a closed loop and always behaves as an ideal gas;
2.) All the processes that make up the cycle are internally reversible.
3.) The combustion process is replaced by a heat-addition process from an external source.
4.) The exhaust process is replaced by a heat-rejection process that restores the working fluid to its initial state.

Fig. (**8.6**) schematically shows the actual operation and air standard assumption for engine analysis. Applying the air standard assumption and combining Equations 13-15, one will reach:

$LHV \cdot \dot{m}_f = \dot{m}_{ex}\left(h_{ex}-h_{ex,77°F}\right) + \dot{W} + \dot{Q}_{loss}$ ⟶ **16**

To further simplify the analysis, the cold-air standard assumption can be made that the working fluid air has a constant specific heat equal to that at room temperature of 25 $^\circ$C or 77 $^\circ$F. Then, the above equation will change to:

$$LHV \cdot \dot{m}_f = \dot{m}_{ex} \cdot C_{p,air@77\,\mathrm{F}}(T_{ex} - 77°\mathrm{F}) + \dot{W} + \dot{Q}_{loss} \qquad \textbf{17}$$

Using Equation 17, one can estimate the four items in the heat balance, including the fuel's combustion energy, heat loss in the exhaust gas, mechanical work, and heat loss from the engine surface. Additionally, the engine efficiency can be expressed by the ratio of the mechanical work and the fuel's energy in the above equation:

$$\eta = \frac{\dot{W}}{LHV \cdot \dot{m}_f} \qquad \textbf{18}$$

8.4. FUELS FOR ENGINES

Engines need fuels to operate and produce work. Fuel physical properties, such as autoignition quality, octane rate, and volatility, are important to their application as engine fuels. Table **8.2** lists several fuels and their properties. In Otto cycle engines, fuel and air are premixed before feeding into the cylinder for combustion, which is ignited at the end of the compression stroke using electrically activated spark plugs. Thus, the fuel must be volatile to be well premixed with air. In the piston compression of the fresh air-fuel mixture in the cylinder, assuming it is adiabatic, the compression work will increase the mixture temperature. The temperature increase is determined by the compression ratio with the maximum temperature reached at the TDC if there is no ignition. In undesirable engine design, this maximum temperature may be higher than the ignition point of the fuel, leading to autoignition and reduction of the engine power output and efficiency. Thus, the fuel's autoignition quality imposes an upper limit for the compression ratio. Further, combustion of the premixed charge in the cylinder is complex, involving initial SI ignition and rapid flame propagation, along with pressure fluctuation. The fuel's property of handling compression during combustion is important to mitigate "knock." For example, in combustion, the high-pressure zone will greatly compress the unburned mixture, usually near the wall or away from the SI, which may cause detonation and engine "knock." Fig. (**S8.1**) (below) is a sketch of combustion in the cylinder near TDC. On the left is 'normal' combustion, where the flame front propagates uniformly. On the right is pre-ignition or 'knock' where local temperatures due to too high a compression ratio cause unwanted spontaneous combustion. In spark-ignited engines, exact control ignition timing is essential for efficient performance. Undesirable spon-

taneous combustion is often accompanied by a knocking sound from the pressure wave, and in the long run, can cause engine damage. "Knock" can be prevented by using better fuels that do not self-combust (usually indicated by a higher octane rating), careful design of the combustion chamber, and employing a knock sensor that triggers the spark plug at the incipient "knock." Engine "knock" is a major source for cylinder degradation.

Gasoline is a popular fuel for Otto cycle engines. In gas stations, several rates of gasoline are available, *e.g.* 87 (regular), 88–90 (midgrade), and 91–94 (premium). The number denotes the octane rating, a measure of a fuel's ability to avoid "knock." A rating of 90 is equivalent to a mixture of 90% iso-octane and 10% heptane in "knock" handling. The higher the octane number, the more resistant the gasoline is to the engine "knock." Use of higher octane fuels also enables higher compression ratios and turbocharging for better efficiencies and larger power output. In the early 20th century, it was found that adding a compound of lead (tetraethyllead, $Pb(C_2H_5)_4$) to gasoline increased the octane rating and mitigated engine "knock." Leaded gasoline was a popular fuel until proven serious health effects of exhausting lead to the atmosphere caused it to be banned in the middle 1970's.

Table 8.2. Typical fuels and their properties.

Fuel	Density (kg/m^3)	Specific energy (MJ/kg)	$CO_2(\frac{Kg}{GJ})$
Hydrogen	0.09 (at STP)	142.2	-
Natural gas	0.777 (at STP)	52.2	50
Gasolines	744	46.5	71
Diesel	836	45.8	69
Ethanol	788	29.8	64

In addition, ethanol, derived from renewable resources (*e.g.*, corns and biomass) using existing alcohol-production infrastructure, can be used as an Otto cycle fuel. Ethanol has substantially reduced chemical energy compared to gasoline and typically burns more cleanly. In practice, it is blended with gasoline for vehicle use. "There are three general categories of ethanol-gasoline blends: E10, E15, and E85. E10 is gasoline with 10% ethanol

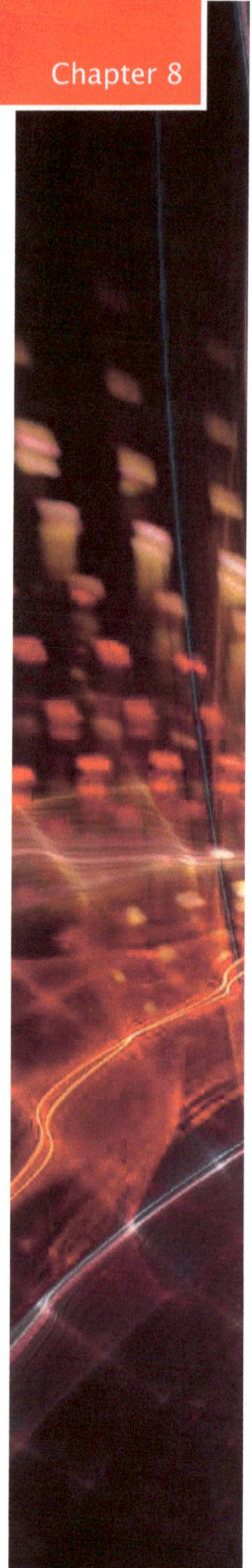

content. E15 is gasoline with 15% ethanol content, and E85 is a fuel that may contain up to 85% fuel ethanol. The ethanol content of most of the motor gasoline sold in the United States does not exceed 10% by volume. Most motor gasoline with more than 10% fuel ethanol content is sold in the Midwest where most ethanol production capacity is located. Gasoline dispensing pumps generally indicate the fuel ethanol content of the gasoline. All gasoline engine vehicles can use E10. Currently, only flex-fuel and light-duty vehicles with the model year of 2001 or newer are approved by the EPA to use E15. Flex-fuel vehicles can use any ethanol-gasoline blends up to E85" [3].

In diesel engines, ambient air is compressed instead of a premixed fuel-air charge; thus the compression ratio r_v can be higher and fuel is much more flexible than Otto cycles. In general, diesel fuel is denser and has more chemical energy per unit of mass than gasoline and thus, diesel engines produce a larger power density than Otto engines. They are often used in heavy-duty applications such as pickup trucks, semi-trucks, trains, farm equipment, and commercial ships. Diesel fuel is usually less volatile; thus its direct use in an ordinary Otto-engine car will cause major problems and operation failure.

8.5. ENGINE TESTING

To measure the engine efficiency of the Otto or diesel cycle, one needs to obtain the power output and the energy input from the fuel, as shown in Equation 18. The former is usually measured using a dynamometer, while the latter can be evaluated by several methods, including flow meters. In this section, the experiment will be introduced as an example to show engine testing. Fig. (**S8.4**) shows the setup of engines and Dyno systems. A sample of engine testing data is shown in Fig. (**8.7**).

8.5.1. Dynamometer

A dynamometer, also called Dyno, is a typical engine-testing device for simultaneously measuring the torque, rotational speed (RPM), and power output. A Dyno system consists of an absorption unit and sensors to measure the torque and RPM. Several types of absorption units have been developed, including eddy current, magnetic powder brake, hysteresis brake, electric generator, and water brake. Table **8.3** lists several types of dynamometers. To measure torque, the dynamometer housing can be mounted, allowing free rotation except as restrained by a torque arm, which is connected to the dyno housing. A weighing scale is then used to dyno housing in attempting to the

dyno housing in attempting to rotate. The torque is then calculated by force multiplied by the length of the torque arm. Fig. (**8.8**) shows the schematic of an electric Dyno system.

	RPM	HP	Torque (ft*lb)	Cosumed fuel	duration [s]	Exhaust Gas Temp [F]	Water (in)
	930	2	11.7	50 ml	159.2	557	0.42
	1718	5	15.6	50 ml	85.4	682	0.96
Gasoline	1936	5.7	15.4	50 ml	71.0	731	1.09
Case #1	2140	6.2	15.4	50 ml	63.9	748	1.2
	2432	7	14.6	50 ml	56.2	776	1.4
	2939	7	12.2	50 ml	49.6	822	1.62
	3415	6.7	10.1	50 ml	50.4	1899	1.74

	RPM	HP	Torque (ft*lb)	Cosumed fuel	duration [s]	Exhaust Gas Temp [F]	Water (in)
	2500	6.5	13.6	50 ml	56.5	1335	1.42
	2675	5.5	10.1	50 ml	67.6	1414	1.22
Gasoline	2585	4.3	8.9	50 ml	75.8	1399	1.02
Case #2	2560	3.1	6.5	50 ml	91.5	1360	0.78
	2545	2.5	5.1	50 ml	112.8	1319	0.63
	2565	1.2	2.6	50 ml	123.0	1280	0.48
	2550	0.3	0.6	50 ml	145.0	1261	0.33

	RPM	HP	Torque (ft*lb)	Intake air speed [m/s]	Exhaust Gas Temp [F]	Initial weight [oz]	Final weight [oz]	Duration [min]
	2116.8	0.5	1.2	8.44	88	240.5	239.7	2
Diesel	2039.4	1	2.6	8.13	107	239.85	239.2	2
Case #1	2043	2	5.2	7.84	148	236.25	235.7	2
	1960	2.5	6.6	7.96	177	235.3	234.85	1
	2081	4	9.9	7.48	256	234.25	233.5	1
	1943	5	13.4	7.61	320	233.35	232.75	1

Fig. (8.7). Sample of Otto and diesel engines testing data. The last column in the Gasoline is the reading from the U-tube manometer of an orifice plate flow meter for air intake with the calibration given by Fig. (**S8.2**) for 0.601 area.

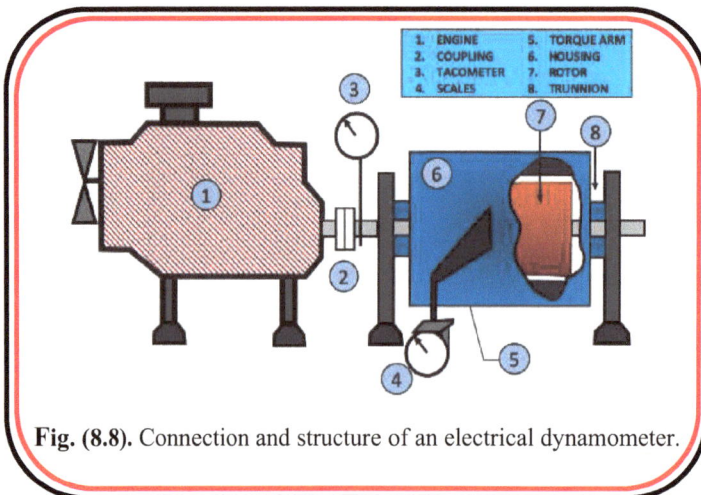

Fig. (8.8). Connection and structure of an electrical dynamometer.

Table 8.3. Types of dynamometers.

Types	Description	Image
Eddy current Dyno	It is the most common absorber in modern chassis dynos and provides a quick load change rate for rapid load settling. Most are air cooled and some require water cooling.	
Powder Dyno	It has a fine magnetic powder placed in the air gap between the rotor and coil. The resulting flux lines create "chains" of metal particulate that are constantly built and broken apart during rotation, creating torque. It is typically limited to low RPM due to heat dissipation.	
Hysteresis Dyno	It uses a magnetic rotor or AlNiCo alloy that is moved through flux lines generated between magnetic pole pieces. The magnetization of the rotor is cycled around its B-H characteristic, dissipating energy.	
Electric motor/ generator	It is a type of adjustable-speed drive with its absorption/driver unit being either an AC or DC motor. The motor can operate as a generator driven by the unit under test.	

As a typical example of dyno, a water brake dynamometer consists of a rotor turning inside a stator with a force gage connected to the stator, as shown in Fig. (**8.9**). As the names imply, the rotor is attached to the crankshaft of the engine with the stator fixed. Water flows through the small passages between the rotor and stator, and viscous stresses transmit the torque from the rotor to the stator. A strain gauge measures the force on the stator at a known radius from the center of the dynamometer, and hence the torque can be obtained. The torque or load on the engine is varied by changing the water flow rate through the Dyno. Too much load will exceed the torque rating of the engine, causing the stall. Thus, it should be changed slowly in operation; abrupt changes may break the crankshaft or other internal parts of the engine. The rotational rate of the crankshaft can be measured by a tachometer, calibrated in revolutions per minute (RPM).

Fig. (8.9). Internal structure, rotor, and stator of a water-brake dynamometer.

8.5.2. Fuel Injection and Air Flow Rate

This section introduces basic methods of measuring the fuel injection rate and the airflow rate, which are important to evaluate the heat balance and understand the air standard assumption.

For fuel injection rate, one can use a stopwatch and a burette with a T-valve to connect with the fuel tank. The principle is the same as that in Chapter 2. By timing a known volume of gasoline from a burette as it flows into the fuel system, one can estimate the fuel consumption rate. The fuel flow into the carburetor is pulsatile due to the fuel float valve opening and closing in response to engine demand. The restriction in the burette damps out the fluctuations. The burette is connected to the engine by a T-valve so that the engine can run on the main gas tank supply until the gas flow rate is measured, whereupon the gas tank is valved off, and the supply from a full burette is opened. As an alternative, one can measure the weight change of the fuel tank in a period of time. To reduce uncertainty, the testing should be longer for smaller consumption rates or lower Hp.

In addition, flow meters can be modified and equipped to measure the average fuel consumption rate. Note that due to the reciprocating nature of the engine, the fuel flow rate fluctuates: for example, during the strokes of the piston expansion and exhaust gas rejection, no fuel is injected. In general, because engine RPM is usually above 1,000, such a short period of fluctuation can be damped out by modifying the flow meter design in the measurement of the average fuel inject rate. In below, a slightly modified orifice plate will be intro-

duced for measuring the average inlet air flow rate.

To measure the airflow rate, one can slightly modify an orifice plate flow meter, which connects the ambient air and the engine airport. In engine operation and air intake, a pressure drop will develop across the orifice. A large tank or bucket can be added as a plenum to damp out the pulsations from the intake strokes of the engine. Calibration then correlates the pressure drop with the airflow rate, as shown in Fig. (**S8.2**) as an example. Turbine meters can also be used to measure the average air intake rate, which directly produces electrical signals.

#Example 3: Calculate the airflow rate, exhaust gas flow rate, and engine efficiency for the first data in case#1 of Figure 8.7. The LHV is given to be 42.5 MJ/kg.

Solution:

From Figure 8.7, the reading for H_2O is 0.42 in.

From Fig. S8.2, 0.42 in of H_2O is corresponding to $\dot{m}_a = 19.5$ lbs/hr.

From Example 2, $\dot{m}_f = 0.792$ kg/hr = 1.75 lbs/hr.

Thus, the exhaust gas flow rate $\dot{m}_{ex} = \dot{m}_f + \dot{m}_a = 21.25$ lbs/hr.

Note that $\dot{m}_f \ll \dot{m}_a$, thus the combustion is lean. Think about the validity of the air standard assumption.

For $\dot{W} = 2\ Hp = 1.49\ kW$, the Otto engine efficiency is given by:

$$\eta = \frac{\dot{W}}{LHV \cdot \dot{m}_f} = \frac{1.49kW}{42.5MJ/kg \cdot 0.792kg/hr} = 16\%$$

This is much smaller than the thermal efficiency $\eta_{th,\,Otto}$ (57.5%) under the same compression ratio.

8.5.3. HP$_{max}$ *versus* RPM

In engine application, the power transmission unit is usually equipped for controlled application of the engine power. In ICE-based vehicles, gearboxes are an essential part of the power transmission. Five-speed gearbox transmission is popular in modern cars. It provides the speed and torque conversions from a rotating power source to another device. To gain more power and torque from an engine in driving such as climbing a mountain or acceleration, the gearbox needs to shift manually or automatically to a higher ratio, permitting more RPM and hence combustion expansion strokes per min in the engine.

In engine testing, power output can be measured at a fixed RPM with the load gradually increased just before the engine stall, *i.e.*, the maximum or peak torque that the engine can produce at a given RPM. Of course, to increase the load, more fuel needs to be added to the engine at the fixed RPM. However, fuel addition can increase load only for complete combustion. In incomplete or fuel-rich combustion, adding more fuel, meanwhile less oxygen, will reduce engine power output under the RPM. Thus, there will be a maximum or peak torque and hence load at each RPM. In the engine testing, it may take a bit of experience with the fuel throttle and dyno load to find the maximum or peak torque. In practice, any load or torque beyond this maximum point requires a shift to another gear ratio in the transmission to enable a higher RPM. So, various RPM will be measured. For a given engine, there is a maxim torque or horsepower (HP$_{max}$) that one can obtain. Thus, the strategy is to increase the RPM in the testing until finding the maximum horsepower. In your car, you can check the maximum power of the model and think about why it is important. The testing data are used to make torque and horsepower versus RPM plot, similar to those seen in car enthusiast magazines. Fig. (**8.10**) shows an example of Hp at the peak torque in Hp versus RPM. Once the engine HP$_{max}$ is reached, friction will dominate and the horsepower will decrease as the RPM increases. Note that the friction work is nearly proportional to RPM, to be introduced in the following sub-section.

In the experiment, i) turn on an engine, ii) set an RPM, iii) gradually increase the load just before the engine stall, iv) conduct measurement of the Hp, torque, fuel injection rate, air injection rate, and exhaust gas temperature under the RPM, v) change to another RPM and repeat iii)-iv) steps.

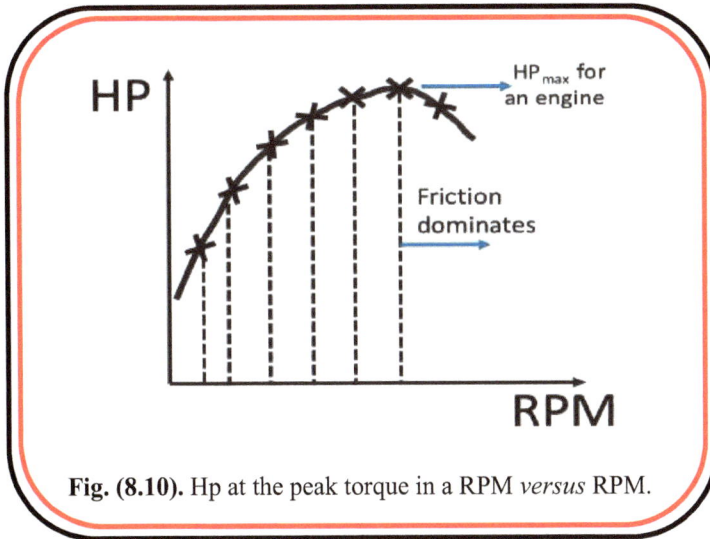

Fig. (8.10). Hp at the peak torque in a RPM *versus* RPM.

8.5.4. Friction Work

In engine operation, the piston will rapidly move in the cylinder at an average speed proportional to the RPM, as seen in Equation 7. The piston is close in physical touch with the cylinder's inner wall to minimize gas leakage in the cylinder, thus a contacting friction force will be present against the piston movement. In addition, the rotating crank arm and any moving components contacting a stationary part will introduce friction, which reduces the brake power output. In a fresh engine, those friction forces can be minimized in the design and by lubrication. For aged engines which may be subject to cylinder erosion and corrosion or lack of engine oils, the friction force can be significant. For example, engine "knock" is a major source of cylinder wall erosion. If we only focus on piston movement, the friction work is then evaluated by the friction force multiplied by the distance the piston travels, which is proportional to the RPM.

To measure the friction work, the engine can be tested at a fixed RPM with the load varied from the maximum down to the minimum that is measurable. In the experiment, i) turn on an engine, ii) set a RPM (*e.g.* around 2,500), iii) set a load, iv) conduct measurement of the Hp, torque, fuel injection rate, air injection rate, and exhaust gas temperature under the load and RPM, v) change to another load and repeat iv) step. At each load, two sets of data will be used for the friction work evaluation, including the fuel injection rate and power output. Ideally, these data will follow a linear trend assuming that the thermal energy addition by combustion is proportional to the fuel injection rate, as shown in Fig. (**8.11**). By linearly extrapolating these data, one will obtain two intercepts with the x and y axes:

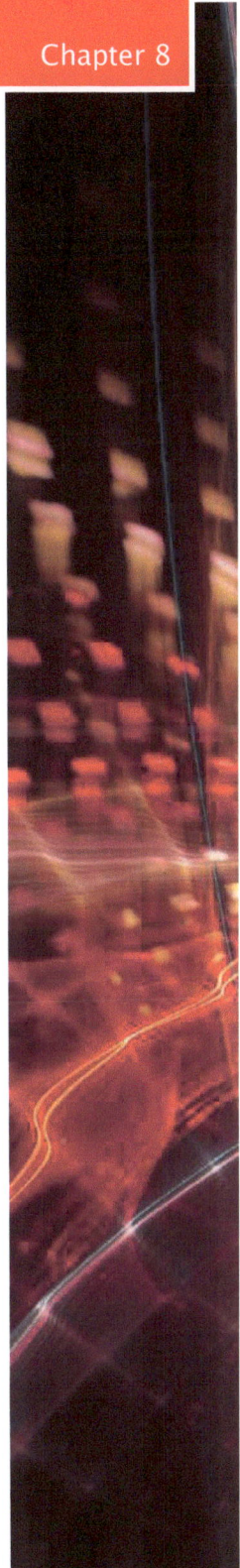

- The y-intercept represents the fuel consumption rate that produces zero brake power. In practice, this intercept at the lowest shift denotes the fuel consumption rate at the idle operation, for example, vehicle operation at a traffic light. The fuel is consumed to just overcome the friction work at the RPM.
- The x-intercept represents the friction work at the RPM. For aged engines with eroded cylinder walls, the friction work will increase, reducing the fuel efficiency.

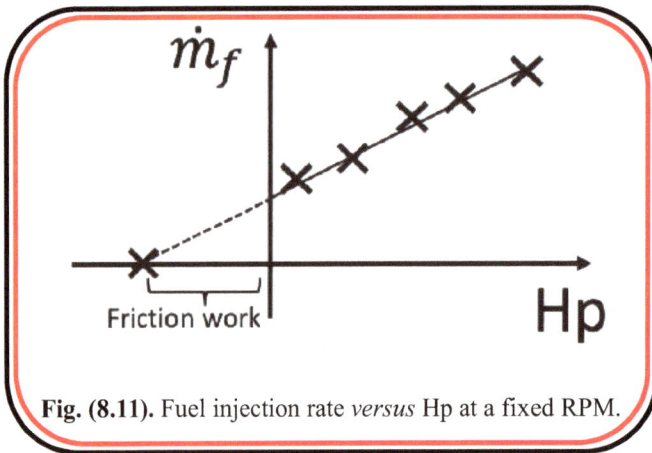

Fig. (8.11). Fuel injection rate *versus* Hp at a fixed RPM.

#Example 4: Calculate the friction power using the first and last data of the case#2 in Figure 8.7.

Solution:

For case#2, the first data give the fuel consumption rate:

$\dot{m}_f = 0.05L/145.0\ s \times 0.001\ m^3/L \times 720\ kg/m^3 = 0.248\ g/s$

at the engine power 0.3 Hp.

The last data give the fuel consumption rate:

$\dot{m}_f = 0.05L/56.5s \times 0.001\ m^3/L \times 720\ kg/m^3 = 0.637\ g/s$

at the engine power 6.5 Hp.

> *Thus, the x-intercept in Figure 8. 11 is calculated to be -3.65 Hp, and the friction power is 3.65 Hp around 2,500-2,550 RPM based on the two data points. In practice, multiple data are obtained for linear fitting.*

For further readings regarding engines and experiment, we refer the interested readers to the references in [4-8].

8.6. QUESTIONS

1. Write the first-law cold-air-standard balance for the engine and calculate the heat losses in Hp and thermal efficiency.

2. Using the sample data of engine testing, plot heat loss and efficiency *versus* (1) engine RPM (for the Otto cycle) and (2) *versus* load (torque).

3. Using the sample data, plot the peak torque and power at a given RPM *versus* RPM. Does peak horsepower occur at peak torque? (for the Otto cycle only).

4. From the runs, a fixed RPM and various loads, make a 'Willian's Line' plot of fuel rate *versus* brake horsepower using the sample data of engine testing. Calculate the friction work.

5. Plot bsfc *versus* bmep for the fixed RPM run. Discuss the observed variation: is there a minimum? (for the Otto cycle only)

6. Sketch two lines in Fig. (**8.11**) for RPM1 and RPM2 with RPM1 = 2xRPM2.

7. Sketch two lines in Fig. (**8.10**) for a small sedan *versus* a SUV.

8. Sketch two lines in Fig. (**8.11**) for an Otto cycle and a diesel cycle engine: note that a diesel engine usually runs under lower RPM, but high HPmax.

9. Check your car model (if you don't have one, pick one of your favorites through online) and find the information regarding gearbox, MPG, and HPmax. Find a way to estimate the fuel efficiency of the car model.

10. Explain how fuel is burned within the cylinder of an internal combustion engine and what happens to mechanical power output as RPM changes.

11. A tank of volume V1 is filled with gas to a pressure of P1. If the gas is allowed to expand into another tank of volume V2 until equilibrium is reached, what will be the pressure in each tank at equilibrium?

12. Suppose you work for an engine manufacturer. A client is complaining that one of the natural gas engines you sold them is running rough. How would you troubleshoot?

13. Describe how to make an Otto cycle engine more efficient.

14. List a few benefits of using Otto cycle engines to replace steam turbines in power plants.

15. Describe how to improve the power output of an Otto cycle engine.

16. Describe how to improve the peak torque of an Otto cycle engine.

REFERENCES

[1]   N. Otto, "Gas-Motor engine," US patent application US, vol. 365701, 1887.

[2]   Y. A. Cengel and M. A. Boles, "Thermodynamics: An engineering approach 6th Editon". The McGraw-Hill Companies, Inc., New York, 2007.

[3]   Available: https://www.eia.gov/

[4]   J.P. Wang and Y. Wang, "Direct numerical simulation of turbulent flames in HCCI-Engines", International Conference on Applied Computational Fluid Dynamics, Beijing, Oct. 2000, P484.

[5]   J.P. Wang, Y. Wang, Y. Umeda, M. Tamura, and Y. Takemoto, Numerical simulation of NOx exhaust of homogeneous charge compression ignition, J. of Environmental and Information Sciences, Yokkaichi Univ., V4, NO. 2, 2000, P1.

[6]   V. Smil, "Prime movers of globalization: The history and impact of diesel engines and gas turbines". MIT press, 2010.

[7]   A. J. Martyr and M. A. Plint, "Engine testing: theory and practice". Elsevier, 2011.

[8]   R. E. Sonntag, C. Borgnakke, G. J. Van Wylen, and S. Van Wyk, Fundamentals of thermodynamics, New York: Wiley, 1998.

Practical Handbook of Thermal Fluid Science, 2023, 160-176

REFRIGERATION

9.1. INTRODUCTION

Refrigerators and air conditioners are popular and important devices in our daily lives. The former provides a low-temperature environment to store food and drink, while the latter reduces the indoor temperature to a comfortable level. In nature, we observe that heat flows from a higher temperature toward a lower temperature. To reverse the heat flow from a low to high temperature, mechanical work needs to be added to enable mechanical refrigeration, by which a low-temperature environment is created for food saving or air conditioning. The most important feature of mechanical refrigeration is its high efficiency, usually called the coefficient of performance (COP): the COP of residential refrigerators is usually above one, and air conditioners can achieve even five. While other refrigeration methods have been developed, based on different physical principles, such as vortex tubes and thermoelectric coolers, their efficiencies, in general, are much less than one, and we use them for special applications where their other characteristics make them attractive.

Prior to the invention of the practical vapor compression mechanical refrigeration equipment by James Harrison in 1851, refrigeration in temperature latitudes was accomplished by harvesting ice in the winter season and storing it in large barns insulated with straw. The ice was preserved for cold food storage in warm seasons, either on the farm or for delivery to residential and commercial units using iceboxes. Trade in ice goes back at least 3,000 years, and with improved insulation and harvesting techniques, became a large trade by the middle of the 1800's. With expanding demand, high-quality ice supplies became more difficult to attain. Around the same time, the first vapor-compression mechanical systems for making ice began to displace harvested ice. In the late 19th century, ice was even used to provide space cooling. The use of ice led to defining a unit in refrigeration, the "ton," also called a refrigeration ton. One ton is defined as the rate of heat transfer that results in the freezing or melting of 1 ton (2,000 lb) of pure ice at 0 °C (32 °F) in 24 hours. It is equal to 12,000 British thermal units per hour. In addition to refrigeration, the cycle can be used for heating at a very high efficiency or COP by using the heat released on the high-temperature or condenser side.

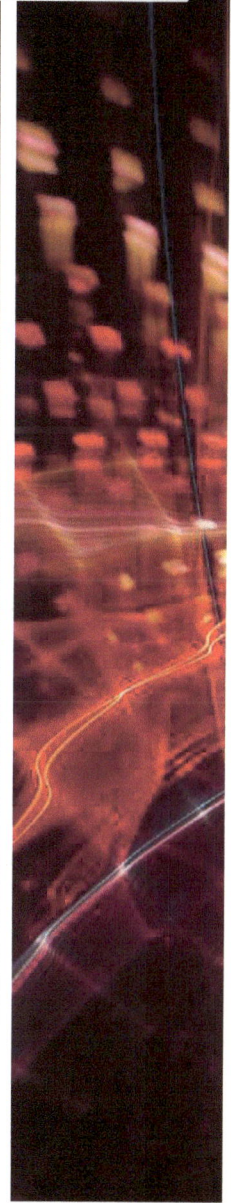

9.2. REFRIGERATION CYCLE

9.2.1. Reversible Carnot Cycle

The fundamental basis of initial analysis and design for mechanical refrigeration is the Carnot cycle, which produces work from two thermal reservoirs with heat flow from T_H to T_L. Since the Carnot cycle is reversible, theoretically, it can be used as a refrigerator as well, which drives a heat flow from T_L to T_H by mechanical work. The vapor compression, heat rejection, and heat addition from the load are all feasible with practical components, *i.e.* a compressor and two heat exchangers, respectively. The difficult remaining part of the Carnot refrigeration cycle is the change of state from a saturated liquid at high pressure to a cold vapor-liquid mixture at low pressure in the evaporator. In principle, this could be done in a power-producing device such as a turbine. Although this is done where large volumes of compressed and liquefied gases are used (*e.g.* oil refineries, compressed natural gas plants, and the like), most refrigeration systems, even large ones, are far too small for this to be practical.

Instead, the typical method of accomplishing the change of state in mechanical refrigeration is to use a simple *throttling device (or valve)* to drop the pressure and accept the loss of efficiency since no power is obtained. A pressure drop occurs across flow restriction such as a partly closed valve due to turbulent dissipation in the vortical eddies downstream of the valve and others. When a saturated liquid passes through an ideal insulated throttling valve, vaporization takes place because the pressure is reduced. Fig. (**9.1**) shows the major components of a refrigeration cycle.

9.2.2. Refrigeration Cycle

An ideal refrigeration cycle consists of two constant-pressure processes with heat transfer, one isentropic process, and one constant-h process, as shown in Figs. (**9.1** and **9.2**):

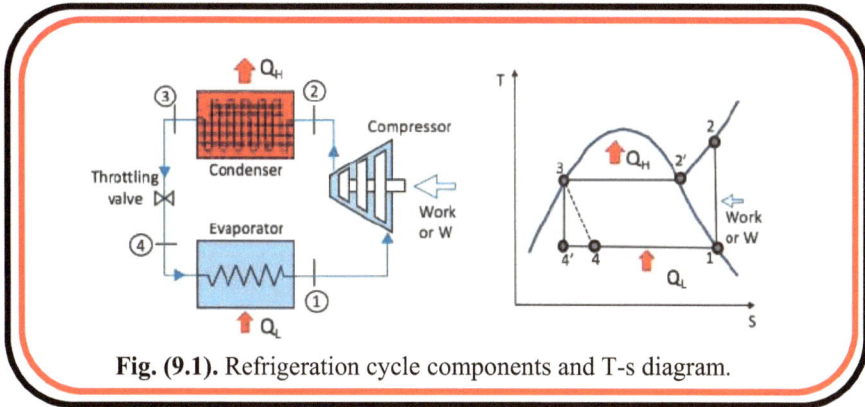

Fig. (9.1). Refrigeration cycle components and T-s diagram.

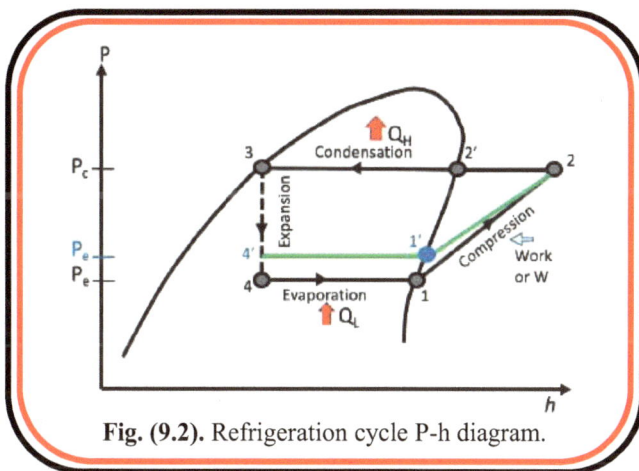

Fig. (9.2). Refrigeration cycle P-h diagram.

1-2 Vapor Compression: Ideal isentropic **compression** of saturated refrigerant vapor to superheated vapor.

2-3 Heat Rejection: Ideal constant-pressure phase change in a **condenser** from the super-heated vapor state to a saturated liquid. Heat must be rejected to the environment. Note that this consists of first *sensible* heat transfer from the super-heated state to saturated vapor followed by *latent* heat transfer from the saturated vapor to saturated liquid.

3-4 Throttling Valve: Isenthalpic **throttling** of the high pressure saturated liquid to a cold mixture at the low evaporator pressure.

4-1 Heat Addition: Ideal constant-pressure evaporation of the cold mixture to saturated vapor in the **evaporator**. Heat addition is from the **load**.

While the whole refrigeration system is a closed cycle, we can analyze each component separately as an open-system device. The analysis for the

compressor is straightforward using the ideal isentropic assumption. The analyses for the condenser and evaporator are based on ideal heat exchangers with no pressure drops.

As mentioned previously, the new item in the mechanical refrigeration cycle is the throttling valve (TV), also called an expansion valve and a thermostatic expansion valve (TEV) if it provides control of the evaporator outlet. The ideal throttling valve is insulated and, of course, has no work interaction with the surroundings. Thus, an open system analysis results in no change in specific enthalpy across the valve. The flow process is *isenthalpic.* The refrigerant entering the valve is a saturated liquid, relatively warm and at the highest pressure in the cycle. The restriction in the throttling valve causes a pressure drop due to fluid viscosity and turbulence. As the pressure drops, some of the saturated liquid evaporates and the temperature decreases per the saturation vapor pressure curve characteristic of the refrigerant. Another way to think of the process is downstream of the valve the vapor has a larger enthalpy than the liquid, and in order for the total enthalpy to remain equal to that of the incoming saturated liquid, the temperature must decrease. While the enthalpy remains constant, there is a large increase in entropy; the process is irreversible. Because of the irreversibility, we always plot the isenthalpic throttling process as a *dashed line* on a **P-h** or other thermodynamic charts, as shown in Fig. (**9.1**) and Fig. (**9.2**). The other three processes are ideally reversible, so we plot their paths with *solid* lines.

For a refrigerator, the heat absorption in the evaporator during the evaporation of the low pressure, low-temperature refrigerant liquid is utilized. For a heat pump, the heat rejection in the condenser as the refrigerant changes from saturated vapor to saturated liquid is utilized.

9.3. COEFFICIENT OF PERFORMANCE (COP)

The thermal efficiency of the mechanical refrigeration cycle is defined as the ratio of what is utilized (rate of the cooling load) to the compressor power, usually called the Coefficient of Performance (COP):

$$COP_{ref} = \frac{Q_L}{W} = \frac{\dot{Q}_L}{\dot{W}} \qquad\qquad 1$$

Where $\dot{Q}_L$ and $\dot{W}$ are the heat transfer rate from the load in the evaporator and the compressor power input, respectively. The COP can be further related to the temperatures of the two thermal reservoirs, as below:

$$COP_{ref} = \frac{\dot{Q}_L}{\dot{Q}_H - \dot{Q}_L} = \frac{1}{\frac{\dot{Q}_H}{\dot{Q}_L} - 1} \xrightarrow{ds = (\frac{\delta Q}{T})_{rev}} \frac{1}{\frac{T_H}{T_L} - 1} \qquad \longrightarrow \qquad 2$$

For a given refrigerant, the heat addition and rejection rates and compressor power can be determined by its enthalpy changes, as shown in Fig. (**9.2**):

$$\dot{Q}_H = (h_3 - h_2) \cdot \dot{m} \qquad \longrightarrow \qquad 3$$

$$\dot{W} = (h_2 - h_1) \cdot \dot{m} \qquad \longrightarrow \qquad 4$$

$$\dot{Q}_L = (h_1 - h_4) \cdot \dot{m} \qquad \longrightarrow \qquad 5$$

Substituting the above three to Equation 1 yields:

$$COP_{ref} = \frac{h_1 - h_4}{h_2 - h_1} \qquad \longrightarrow \qquad 6$$

As to heat pumps, the COP can be expressed by using $\dot{Q}_H$ as useful heating power:

$$COP_{HP} = \frac{\dot{Q}_H}{W} = \frac{\dot{Q}_H}{\dot{Q}_H - \dot{Q}_L} = 1 - \frac{1}{1 - \frac{T_L}{T_H}} = COP_{ref} + 1 \qquad \longrightarrow \qquad 7$$

Thus, heat pumps operate under an efficiency higher than their counterpart refrigerators.

9.4. REFRIGERANT

Refrigerant is the working fluid in refrigerators or heat pumps. In the four main processes of the refrigeration cycle, the refrigerant changes phase from vapor to liquid in the condenser and then back to vapor in

the evaporator with its latent heat released and absorbed, respectively. In refrigerators, heat absorption during refrigerant vaporization is utilized. In heat pumps, the heat release during refrigerant condensation is utilized.

Selection of refrigerants is important to residential applications. The ideal refrigerant is non-corrosive, non-toxic, not flammable, and has a boiling point below the temperature target. In the 20th century, fluorocarbons, especially chlorofluorocarbons, were widely used as they had all the desirable characteristics listed. Developed in the 1930's, these refrigerants soon dominated the market.

Unfortunately for the refrigeration industry and fortunately for the living things on the planet, researchers at the University of California, Irvine and elsewhere, discovered and proved to the world that chlorinated refrigerants damaged the earth's ozone layer, which protects life on the surface from harmful ultraviolet radiation. An international treaty (the Montreal Protocol) resulted in banning the production of these refrigerants over a period of years. The treaty has been successful, amended, and expanded many times, and the damage to the ozone layer appears to be reversible to 1980 levels by 2070.

Other common refrigerants used in various applications are ammonia, sulfur dioxide and non-halogenated hydrocarbons, such as Isobutane (R-600a) and propane (R-290), which do not deplete ozone and have no or only a few global warming effects. They are widely used in air-conditioning systems for buildings, sport and leisure facilities, the chemical/pharmaceutical industry, the automotive industry and almost all the food industry (production, storage, marine shipping, and retailing). Table **9.1** shows the typical refrigerants and their relevance to ozone depletion.

Table 9.1. Typical refrigerants.

Type	Name	Chemical Formula	Global Warming Potential 100-year	Semi-Empirical Ozone Depletion Potential
CFC	R-13	$CClF_3$	13900	1
PCC	R-10	CCl_4	1400	0.73
H	R-13B1	$CBrF_3$	6290	16
PFC	R-14	CF_4	7390	0
HCC	R-40	CH_3Cl	13	0.02
HFC	R-134a	$C_2H_2F_4$	1300	0
HCFC	R-123	$C_2HF_3Cl_2$	79	0.02

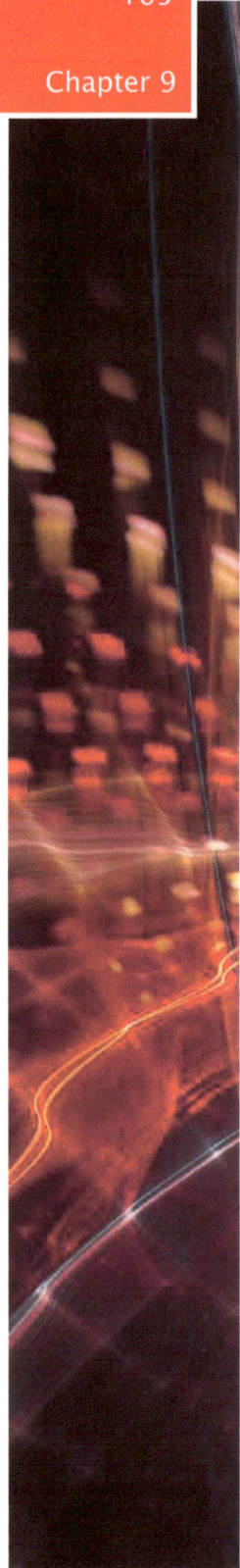

9.5. RANQUE-HILSCH VORTEX TUBE

The vortex tube, also known as Ranque-Hilsch vortex tube, is a refrigeration device that uses compressed gas to produce hot and cold streams, as shown in Fig. (**9.3**). It has no mechanical moving parts and uses only compressed gas as energy input. Compressed air at room temperature is injected into the vortex tube through a nozzle, which forces the gas to spin and transport horizontally in the tube. The spinning air travels down toward one end of the tube, where a special valve is placed to allow only the gas near the tube wall to escape and divert that near the centerline back as a second or inner vortex. The angular momentum is conserved in the tube; thus the inner vortex indicates a rotation with a larger angular velocity. The inner vortex loses angular momentum by interacting with the outer vortex and the kinetic energy is lost in the outer vortex. The inner gas pressure is smaller than the outer vortex due to the spinning; thus the process involves both gas compression and expansion. The outer vortex becomes warm while the inner vortex is cooled. The gas emerging from the hot end can reach a temperature of 200 °C and that from the cold end can reach -50 °C. The COP for a vortex tube refrigerator varies from 0.42 to 0.83 for L/D =10 and from 0.55 to 0.82 for L/D=30 [1]. Commercial vortex tubes for industrial purposes can produce a temperature drop up to about 70 °C. Vortex tubes are widely used for cooling of cutting tools, such as lathes, mills, and CNC machines, by using compressed air available in machine shops. This vortex tube cooling significantly reduces the need for liquid coolants and refrigerants, which can be messy, expensive, and environmentally hazardous. Table **9.2** summarizes several refrigeration methods.

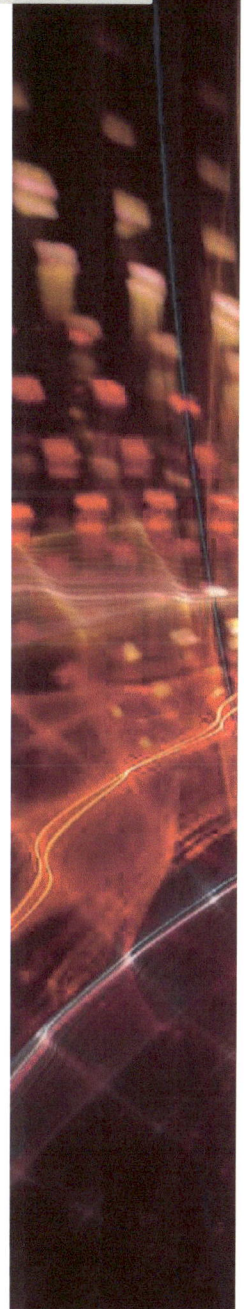

Table 9.2. Refrigeration Methods.

Types	Function	Design
Vapor Compression	Refrigerant undergoes phase changes, and heat is absorbed in the evaporator for cooling.	
Vortex Tube	It separates high-pressure gas into a cool stream at one end and a hot stream at the other.	
Absorption Refrigeration	It employs a heat source to provide the energy and to drive a cooling process by using two coolants, with one for evaporative cooling and absorption into the other. Heat is needed to reset the two coolants to their initial states.	

(Table 9.2) cont.....

Types	Function	Design
Adsorption refrigeration	Similar to absorption refrigeration, but in adsorption refrigeration, the working fluid molecules adsorb onto the surfaces of solid sorbents instead of dissolving into a liquid solution.	
Thermoelectric cooling	A solid-state active heat pump that transfers heat from one side of the device to the other, with consumption of electrical energy	
Magnetic refrigeration	Exposing a material to a changing magnetic field causes material temperature change.	

(Table 9.2) cont....

Types	Function	Design
Thermoacoustic refrigeration	It uses high-amplitude sound waves to pump heat from one place to another.	
Elastrocaloric refrigeration	A novel alternative solid-state cooling technology, based on the latent heat associated with the martensitic phase transformation process.	
Air cycle machine	It is used in jet engine-powered aircraft by passing air through and into the airplane as the refrigerant.	

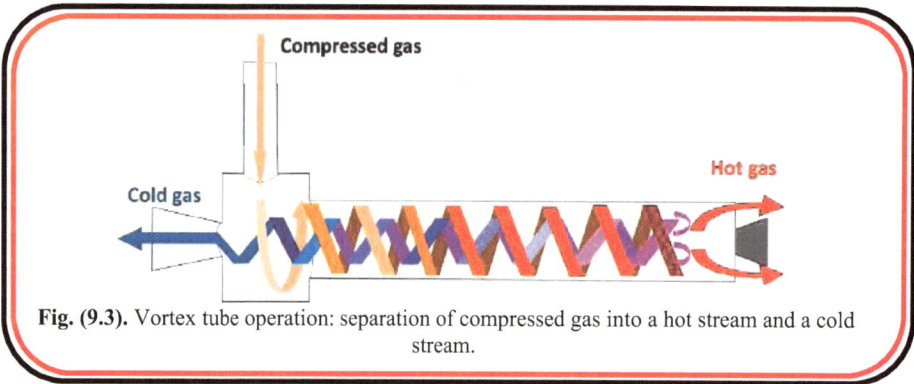

Fig. (9.3). Vortex tube operation: separation of compressed gas into a hot stream and a cold stream.

9.6. REFRIGERATION EXPERIMENT

9.6.1. Experimental Apparatus and Testing

To understand the realistic operation of the mechanical vapor-compression refrigeration cycle, temperature and pressure of the major components and compressor power will be measured. These data provide useful information

for calculating the COP and cooling load, along with drawing cycles in the P-h diagrams. A sample of experimental data based on an R134a system is shown in Table **9.2**, including temperatures, pressures, refrigerant flow rates, and compressor powers. The experimental apparatus is shown in Figs. (**S9.1** and **S9.2**), which consists of the four major refrigeration components and other components, including a slight glass, filter, receiver, accumulator, low and high pressure cut-out relays, and an evaporator thermostat. It can operate as a heat pump if the heat in the condenser is used; thus we can also evaluate the COP and heating load of the heat pump. Because the compressor works with pure vapor, an accumulator is equipped before the compressor to enable vapor-liquid separation by gravity and permit the only vapor to enter the compressor. Similarly, a receiver is equipped before the valve to permit only liquid refrigerant to enter the valve. The operating principle is the same as the deaerator in a Rankin-cycle steam turbine, as discussed in Chapter 3. The refrigerant mass flow rate through the system is measured *via* a rotameter and the electrical power of the compressor is monitored by a digital power meter. Thermocouples and pressure gauges are placed at strategic locations throughout the cycle for temperature and pressure measurement. Transparent tubes are used to view the refrigerant state at several locations.

In the refrigeration industry, unique pressure gauge scales are used. Because the refrigerant is either saturated liquid or vapor at three of the four key points in the cycle (see Fig. **9.1**), the refrigerant temperature is known when the pressure is available. Therefore, a temperature scale corresponding to the vapor pressure can be printed on the gauge scale alongside the pressure scale. Under the saturation restriction, a pressure gauge can be used to measure temperature. Note that the pressure gauges may have various temperature scales for different refrigerants.

In the experiment, i) take the measurements of temperatures, pressures, and refrigerant flow rate before starting the refrigerator, ii) turn on the compressor and set the manual valve to a position, iii) after temperatures are stabilized, take the measurements of temperatures, pressures, compressor power and refrigerant flow rate, iv) set the manual valve to another position and observe the temperature change, v) repeat iii)-iv) steps for more valve positions and refrigeration cycles.

Table 9.3. Sample of Experimental Data.

Variable / Run #	1	2	3	4	5	6	7	8
TEMPERATURES (F)								
Ambient	70	71	71	71	71	70	70	70
1-Comp_out	68	140	161	162	162	159	159	155
2-Comp_in	69	59	51	60	64	67	68	69
3-Cond_out	69	124	131	124	114	107	104	96
4-Cond_in	69	140	158	158	157	152	149	142
5-Exp V in	68	109	118	114	104	101	95	90
6-Evap out	68	48	40	58	60	61	62	63
7-Exp V out	67	38	42	35	26	17	13	4
Evap air out	70	57	61	60	58	60	66	65
Cond air out	70	84	89	82	82	82	78	78
PRESSURES (psi, F)								
Comp out (red)	65,70	200	225	205	176	155	150	133
Cond out (red)	65,70	190	210	200	170	150	145	129
Exp V out (blue)	65,68	33	39	32	22	15.5	13	6.9
Comp in (blue)	65,68	20	25	20	13	8	5	1
ROTAMETER (LPM)	0	0.355	0.41	0.345	0.268	0.21	0.188	0.143
COMPRESSOR (W)	off	630	735	665	580	506	471	405
EXPANSION VALVE								
TXV	on	on	off	off	off	off	off	off
CTV	off	off	off	off	off	off	off	off
AXV	off	off	off	off	off	off	off	off
manual	off	off	on	on	on	on	on	on

9.6.2. Cycles in P-h Diagram

Using the temperature and pressure data, one can plot the refrigeration cycle in the P-h chart of the refrigerant, as shown in Fig. (**S9.3**). Note that the data are from the real operation of a refrigeration system; thus irreversibility's impacts are taken into account, which causes deviations from the ideal paths. There are several major sources of irreversibility, including:

1.) Turbulence in the throttling valve

2.) Viscous flows in the pipes

3.) Irreversible process in vapor compression

In 1.), the ideal path is constant enthalpy from high pressure to low pressure. Because the valve is insulated, one can also think of the cooling effect as resulting from the liquid itself supplying the required latent heat of vaporization. The loss of the latent heat from the liquid results in the mixture temperature decreasing. This all happens very fast and irreversibly in the valve. The ideal insulated throttling valve is characterized by a constant enthalpy process with however an increase in entropy. As stated on the P-h chart, we draw the conceptual reversible parts of the cycle with solid path lines, whereas the throttling process is always shown with a *dashed* line to emphasize the irreversibility. In addition, though insulated, there is a certain amount of heat exchange with the surrounding in the valve. In the case of net heat addition, the enthalpy will increase. One can touch the valve surface and observe its temperature variation.

In 2.), the heat exchanges are assumed under constant pressure during the phase change of the refrigerant. Thus, in ideal paths, two horizontal lines will be drawn in the evaporation and condensation processes. In real operation, as shown in Figs. (**S9.1** and **S9.2**), the evaporator and condenser are pipe networks, which refrigerant flows through. In a pipe flow, a pressure drop will develop to overcome the viscous force imposed by the pipe wall. As a result, the pressure will decrease along the flow direction.

In 3.), ideal compression in a compressor is assumed to be isentropic, *i.e.* reversible and adiabatic. Actual compression always involves irreversibility, such as the conversion of electrical or kinetic energy to heat and heat exchange with the surrounding. The isentropic efficiency of a compressor is defined as the ratio of the work input in an isentropic process to the work input in the actual process under the same inlet and exit pressures:

$$\eta_c = \frac{Isentropic\ Compressor\ Work}{Real\ Compressor\ Work} = \frac{W_S}{W_{real}} = \frac{h_{2s}-h_1}{h_{2r}-h_1} \qquad \longrightarrow \qquad 8$$

where:

- h_1 is the specific enthalpy of the gas at the entrance
- h_{2r} is the specific enthalpy of the gas at the exit for real process
- h_{2s} is the specific enthalpy of the gas at the exit for the isentropic process

In practice, the entropy will increase due to irreversibility.

9.6.3 COP and Refrigeration Loading

The real COP of each cycle can be calculated using the enthalpy values of several points and the compression power consumption, as shown in Equation 6. The enthalpy values can be obtained using the P-h chart or thermodynamic table of the specific refrigerant based on the pressure and temperature at four typical locations. To improve the COP, one can reduce the temperature difference between the evaporator and condenser, as shown by Equation 6 and Fig. (**9.2**).

The cooling and heating loads can be calculated using the temperature, pressure, and mass flow rate of the refrigerant. The former is the rate of Q_L on the evaporator side and the latter is the rate of Q_H on the condenser side. They measure the heat removal and heating capability of the refrigerator and heat pump, respectively. They are important to the practical installation of air conditioners, refrigerators, and heat pumps to meet specific residential or industrial demands. For refrigeration, the unique unit for the load is the "**ton**" (of refrigeration). As discussed at the beginning of this chapter, it was based upon one ton of ice melting in 24 hours and is equal to 12,000 Btu/h (about 3½ kW). To improve the refrigerator or heat pump loading, one can reduce the temperature difference between the evaporator and condenser or increase the refrigerant flow rate.

Under a fixed mass flow rate and condenser temperature, a lower evaporator temperature can be achieved by reducing the evaporator pressure. However, this operation reduces the refrigeration loading and COP.

On the other hand, the evaporator temperature can be raised by lifting the throttling valve, which increases the evaporator pressure. In early residential refrigerators, defrosting the ice on the inner wall of a freezer chamber was usually done manually. Whenever the freezer is open and outside air enters, moisture will be brought into the freezer. Note that even dry air under room temperature, as long as its relative humidity (RH) is not zero, contains a certain amount of water vapor. The moisture becomes frost at the freezer wall, which builds over time and greatly reduces the interior space and refrigerator efficiency. Before any auto-defrost functions are implemented, manual defrosting had to be done regularly to remove the ice stick to the wall, which can be a difficult and time-consuming task. Nowadays, residential refrigerators are equipped with auto-defrost, which can be achieved by temporally raising the evaporator temperature above 0 °C to melt the ice on the wall in a short period of time.

 # Example 1: in run#2 of Table 9.3, calculate the mass flow rate of the refrigerant (R-134a) in kg/s, refrigeration load and COP.

Solution:

From Table 9.3, the volume flow rate $\dot{V}$=0.355 liter/min or 5.92×10^{-6} m^3/s

The density of R-134a is 76.2 lbs/ft³ or 1220 kg/m³ under 70 °F, then,

$$\dot{m} = \rho \cdot \dot{V} = 1,220 \ kg/m^3 \times 5.92\times10^{-6} \ m^3/s = 0.00722 \ kg/s.$$

One can then find the enthalpies (h) at different points of the refrigeration from R134a thermodynamics table using the pressure and temperature data to calculate the COP and refrigeration load.

For further readings regarding refrigeration and experiment, we refer the interested readers to the references in [2-6].

9.7. QUESTIONS

1. From the sample of the experimental data, please indicate the uncertainty of each measurement.

2. Compare the recorded temperatures from the four pressure gauges to the corresponding temperatures from the thermocouples. Which should agree? Do they? If not, why not?

3. Calculate the load for each run in Tons and the COPs. Comment on the variation of the COPs.

4. Compare and comment on the actual final state of the compressor outlet to that if the compression was isentropic.

5. What are the effects of irreversibility in the compressor, condenser, and evaporator on the cycle? Sketch these on a P-h diagram together with the ideal cycle.

6. Draw the cycles in P-h charts. Clearly show the measured state points and process path.

7. Explain the role of gravity in the experimental system. Could it run upside down?

8. You most likely observed water condensation and possibly ice on the throttle valve tubing and the evaporator from the humid ambient air. Do you observe the same in your refrigerator or the air conditioner in a car? How is the condensation and ice problem avoided in modern home refrigerators and car air conditioners? What happens to the condensed water?

9. Using P-h diagram to show how the auto-defrost function works.

10. Explain physically how the receiver or accumulator works using a schematic.

11. For irreversible compressors, do we need to modify the COP expression in Equation 6? Why or why not?

12. Please use the thermodynamic properties of water to derive and calculate how much kW is for a **Ton** loading.

13. For one **Ton** loading between 20 $^{\circ}$C and -10 $^{\circ}$C, please estimate how much electricity you need to pay, assuming the COP is the theoretical or half of the theoretical one! The electricity is \$0.30/kWh in the daytime and \$0.10/kWh at night. Please comment on the two values and the need to set your refrigerator to run at night!

14. If you use the Vortex tube for refrigeration, please recalculate 13 by considering a COP of 0.8.

15. Your house does not have an air conditioner. Can you cool the room by opening the refrigerator door instead? Explain.

REFERENCES

[1]   U. Behera *et al.*, "CFD analysis and experimental investigations towards optimizing the parameters of Ranque–Hilsch vortex tube,"International Journal of Heat and Mass Transfer, vol. 48, no. 10, pp. 1961-1973, 2005.

[2]   R. E. Sonntag, C. Borgnakke, G. J. Van Wylen, and S. Van Wyk, "Fundamentals of thermodynamics", New York: Wiley, 1998.

[3]   Y. A. Cengel and M. A. Boles, "Thermodynamics: An engineering approach 6th Editon". The McGraw-Hill Companies, Inc., New York, 2007.

[4]   R. P. Mullin, "What can be learned from DuPont and the Freon ban: A case study". Journal of Business Ethics, vol. 40(3), pp. 207-218, 2002.

[5]   F. S. Rowland, "Stratospheric ozone depletion. In Twenty Years of Ozone Decline" (pp. 23-66). Springer, Dordrecht, 2009.

[6]   J.M. Calm and G.C. Hourahan, "Refrigerant data update". Hpac Engineering, 79(1), 50-64, 2007.

REPORT PREPARATION

10.1. OVERVIEW

To formally document a laboratory experiment, a written technical report is prepared. However, it should be noted that detailed lab notes compiled during the experiment are essential, so they can be used later in the full written report.

The overall goal for any laboratory report is to summarize an experiment efficiently for an intended audience. Therefore, knowing who will read the laboratory report (*e.g.* researchers in a similar field in a technical journal, business leaders in a company as an internal white paper, or the general public for mass consumption) will dictate the style, the tone and the level of technical jargon needed in the written work. After identifying the target audience, the lab report should explain the purpose of the experiment, the research questions answered, the experimental procedures, the data observed, the analysis method and the results.

10.2. REPORT FORMAT AND CONTENTS

In general, the basic components in a typical lab report can be organized into the following outline:

I.   Abstract or Summary

II.   Introduction

III. Methods

IV.   Results and Discussions

V.   Conclusions

VI.   References

The abstract or summary is a short overview of the entire lab experiment placed at the beginning of the full lab report. It provides enough detail so the reader can quickly assess if the full report is interesting enough to warrant further attention. Following the abstract is the introduction that describes existing

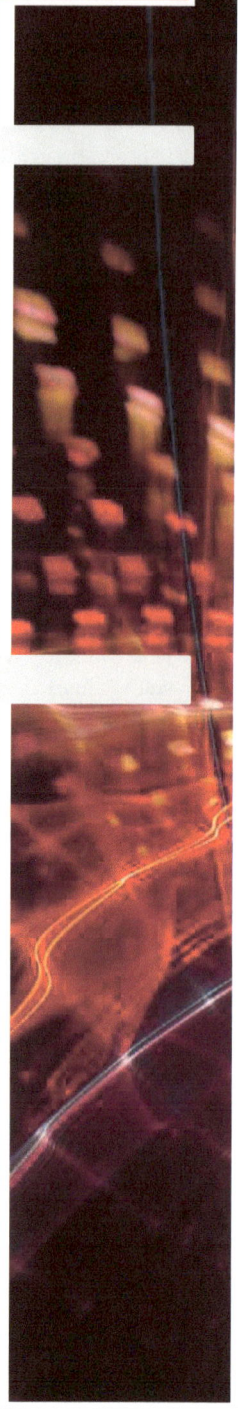

literature and research, current limitation and challenges, and research questions to be answered in the current report. The methods section will describe the experiment in detail, including mathematical equations, apparatus, and procedures. The results and discussion section will document the data, the analysis, and the relevant figures, along with in-depth discussions on how the experimental data can be interpreted. In the end, the conclusion section highlights the major findings from the lab experiment. Of note, as all experiments will involve data measurements and calculations, careful attention must be paid to proper dimension, unit, and number formatting, as well as observing any graphing and charting etiquette.

The following are the detailed description for each lab report component:

I. Abstract or Summary

The objective of this section is to provide an overview of the experiment in a very short, succinct format without mathematical equations, tables, figures, references, abbreviations, and acronyms. A concise and factual abstract is desirable, and in some cases, there is a maximum word count depending on the intended audience, *e.g.* a journal publisher, technical organization or academic institution. The main content in an abstract includes the purpose of the work, what was done, the brief procedures, and main conclusions or key findings. This section should provide the readers an idea about the subject and main results covered in the report, enabling a quick decision whether to proceed reading the full report or not. A well-written abstract or summary attracts attention and encourages readers to examine the rest of the report in detail.

II. Introduction

The objective of this section is to provide a context for the experimental work and methods. A review of previous work on the same subject should be included with critics or comments, followed by highlighting inadequacies, challenges, and limitations. Identifying major problems with the current state of knowledge in the field will help the readers understand the relevance and the broader impacts. The motivation and a brief introduction of the experimental work in the report at the end of this section will address the main challenges and the key research questions.

III. Methods

The objective of this section is to describe the applicable theoretical concepts, mathematical equations, and experimental methods applicable for the experi-

ment. This section enables readers to understand relevant theories in detail and verify the validity of the experiment, and aids in the interpretation of the experimental data.

To describe the theories and equations, a standard format with consistent symbols should follow to facilitate readers' understanding, such as each equation per line followed by succinct descriptions defining all the variables. Each formula should be properly labeled with a number for readers to keep track of when the equation is cited in a later section.

To describe the experimental method, the main apparatus and procedures should be given and listed in detail. Sufficient information must be provided for readers to understand what equipment was used and what was done step by step. In some cases, major specific apparatus brand, model, and precision limits should be given. A flow chart or tabulated procedure can be added to visually show the experimental steps.

IV. Results and Discussion

The objective of this section is to present the results in a style and form meeting the needs of the intended readers. Tabular and graphical presentations are usually used to show important results and findings clearly. A discussion should accompany each figure or table to explain how to understand the graphic and how the results can be interpreted and identify any similarities, differences, or trends to be seen. Depending on the guidelines provided by a journal publisher, technical organization or academic institution, the standard format of figures and tables should be used to facilitate the understanding and avoid any confusion.

V. Conclusions

This section aims to discuss the results and important findings and then draw a conclusion. Major quantitative results are often located in this section. The conclusion is as important as the abstract or summary because most readers first go through these two sections before deciding to read the rest of the report. In some cases, suggestions and future works can be outlined at the end of this section.

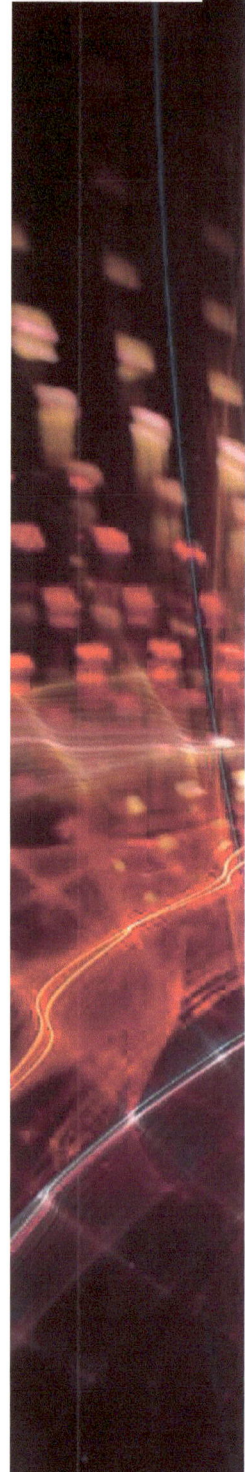

VI. References

Every reference cited in the lab report needs to be present in this section, except the Abstract, where any references should be given in full at the place noted in the Abstract. Unpublished results and personal communications can be cited by using either 'Unpublished results' or 'Personal communication'. For Web references, the full URL should be given and the date when the reference was last accessed. For Data references, they should include the following author name(s), dataset title, data repository, version (where available), year, and persistent global identifier.

Several standard citation formats are available. Depending on the journal publisher, technical organization, or academic institution, the citation style adopted in the reference section can vary. For example, Institute of Electrical and Electronics Engineers (IEEE) has its own guidelines for citations. More examples are given below for a patent, journal paper and book, respectively:

Wang, Yun. "Thermal Devices For Controlling Heat Transfer." U.S. Patent Application No. 14/619,781.

Wang, Y., Chen, K. S., Mishler, J., Cho, S. C., & Adroher, X. C. (2011). A review of polymer electrolyte membrane fuel cells: Technology, applications, and needs on fundamental research. Applied Energy, 88(4), 981-1007.

Mettam GR, Adams LB. How to prepare an electronic version of your article. In: Jones BS, Smith RZ, editors. Introduction to the electronic age, New York: E-Publishing Inc; 2009, p. 281–304.

No matter what citation style is adopted, the format should be consistent throughout the entire report.

10.3. DIMENSIONS AND UNITS

10.3.1. Definitions

Proper unit usage is an essential part of credible and quality engineering analysis. Inconsistent unit usage can lead to embarrassing and disastrous results. In 1999, a NASA spacecraft was lost due to inconsistent units. Afterward, "The peer review preliminary findings indicate that one team used English units (*e.g.*, inches, feet and pounds) while the other used metric units for key spacecraft operation. This information was critical to the maneuvers required to place the spacecraft in the proper Mars orbit" [1].

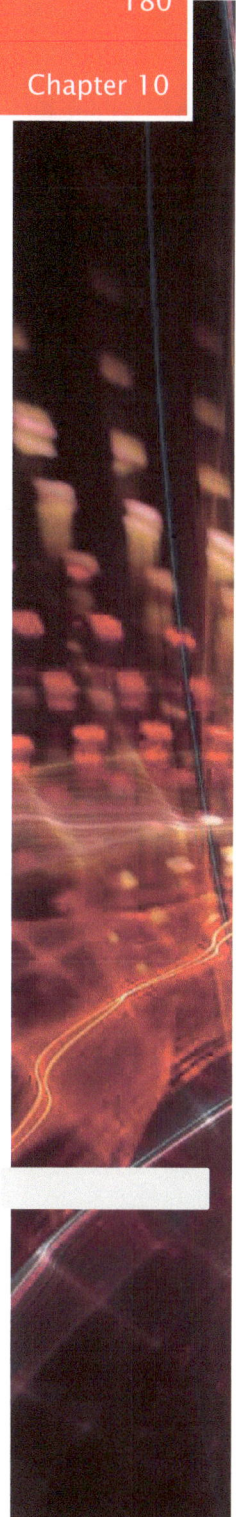

Practical Handbook of Thermal Fluid Science

In a lab report, *dimensions* are used to describe physical variables; *units* are the quantitative implementations of dimensions in a system. *Basic or primary* dimensions are defined, from which *secondary* dimensions are formulated. A simple example is a length:

Variable	Dimension	Unit	
		Metric	English
length	L	meter [m]	foot[ft]

Length is a primary dimension. Secondary dimensions can be formulated from it. For example, the dimension of velocity is LT^{-1} with two possible units ms^{-1} or $ft\ s^{-1}$, where T denotes time.

Dimensional *homogeneity* is required in equations. All terms in an equation must have the same dimensions. Of course, if they have the same dimensions, then they will have the same units in a given system. Also, note that the argument of an exponential or logarithmic operator must have *no* dimensions.

10.3.2. Unit Conversions

The easiest way to manage units is to treat them like algebraic quantities. It is good practice to show all units and conversions explicitly in the sample calculations. Conversions are treated as unitary fractions, which can be used to change from one unit to another. A simple example is:

$$12\ inches = 1\ foot \qquad\qquad\qquad\qquad 1$$

$$1 = \frac{12in}{1ft} \qquad\qquad\qquad\qquad 2$$

$$1 = \frac{1ft}{12in} \qquad\qquad\qquad\qquad 3$$

The unitary fractions can be used for conversion. A length of 29 inches is equal to:

$$? = 29in \cdot \frac{1ft}{12in} = 2.4ft \qquad\qquad\qquad\qquad 4$$

Strikeouts are used to show the algebraic unit cancellations, which will be helpful for future error tracking.

Unit conversions can be readily extended to express answers in terms of secondary units. Using Force as an example, from Newton's law of motion, force, F, equals mass times acceleration. Acceleration is a secondary quantity of dimensions $L\,T^{-2}$. Force is, therefore, another secondary quantity of dimensions $M\,L\,T^{-2}$. In metric units (SI or System International), this secondary group of units is defined to be newton (N). One newton is equal to $1\ kg\ m\ s^{-2}$. Note that when units are named after well-known scientists, the person's name as a unit is not capitalized, but the first letter in the unit abbreviation is, *e.g.*, "newton" [N] and "pascal" [Pa].

To illustrate secondary units in the SI system, the following example is considered: a square block of 1 m on a side with a mass 10 kg accelerates at 1 $m\ s^{-2}$ into a wall. What is the impact force?

$$F = ma \qquad\qquad\qquad\qquad\qquad\qquad 5$$

$$F = 10kg \cdot 1\frac{m}{s^2} \cdot \frac{N\,s^2}{kg\,m} \qquad\qquad\qquad 6$$

$$F = 10N = \frac{1ft}{12in} \qquad\qquad\qquad\qquad 7$$

To extend the above example, one can find the pressure on the wall. Pressure is defined as the normal force per unit area, and the SI unit of pressure is pascal [Pa], which is equal to $1\ N\ m^{-2}$.

$$P = \frac{F}{A} \qquad\qquad\qquad\qquad\qquad\qquad 8$$

$$P = \frac{10N}{m^2} \cdot \frac{Pa\,m^2}{N} \qquad\qquad\qquad\qquad 9$$

$$P = 10Pa \qquad\qquad\qquad\qquad\qquad\qquad 10$$

10.4. SIGNIFICANT FIGURES

The significant figures (SF) of a number are digits that are meaningful, reflecting its measurement resolution. It includes all digits except all leading zeros. In data analysis and report preparation, one will take measurements in experiments and calculate various results from them. The final results are only as accurate and precise as the input measurements allow. Accuracy is determined by the comparison of measurement to an accepted standard. Precision is the degree of granularity of the measurement; high accuracy requires high precision, but high precision does not imply high accuracy. With the advent of digital displays and computers, the distinction between the two has blurred. For example, a gasoline pump in a gas station displays the volume of gas pumped. The volume is given to 0.001 gallons, which usually doesn't represent the accuracy of the measurement. After all, one is paying real accurate dollars and cents based on the indicated volume. A useful website for significant figures is Ref. [2]. Calculations done in Excel can often produce incorrect SFs. Thus, one should properly set the Format Cells option in Excel to specify the number of SFs.

 # Example 1: How many SFs in the voltage measurement $e=0.010$ V, 0.011 V, and 0.01 V, respectively? In a thermocouple with $K=200\ ^{\circ}C/V$, write the temperature readings with correct SFs ($T=Ke$).

Solution:

e (V)	SF	T ($^{\circ}C$)
0.010	2	2.0
0.0110	3	2.20
0.01	1	2

An Excel example is the calculation of the area of a circle ($A = pi \times R^2$) from the radius given to different SFs, as shown in Table **10.1**. It also shows the difference between using Excel's built-in value for pi, written as fx = pi()$\times R^2$ in the Excel formula window and typing in various approximations. Note that when pi () is used in Excel, the number of significant figures in the answer is greater when compared to using 3.14, 3.1416, *etc.* In addition, the use of a known constant or conversion factor should not change the number of significant figures in the answer. When the average of a number of trials is reported, it is acceptable to report the answer to one more significant figure than in the individual trial, *e.g.*, the average of 2 and 3 is 2.5.

Table 10.1. Formatting numbers in Excel.

Radius	SF	Area [pi()]	Area [3.14]	Area [3.1416]	E format	Decimal
2	1	12.56637	12.56	12.5664	1.E+01	10
2.0	2	12.56637	12.56	12.5664	1.3E+01	13
2.00	3	12.56637	12.56	12.5664	1.26E+01	12.6
2.000	4	12.56637	12.56	12.5664	1.257E+01	12.57
2.0000	5	12.56637	12.56	12.5664	1.2566E+01	12.566
2.00000	6	12.56637	12.56	12.5664	1.25664E+01	12.5664
2.000000	7	12.56637	12.56	12.5664	1.256637E+01	12.56637

 # Example 2: Calculate the average and standard deviation for the data set using correct SFs:

1.00	2.00	3.00	4.00	5.00	6.00	7.00	8.00	9.00

Solution:

	Excel	*Correct*
Average	5.00	5.000 (It's OK to have one more SF)
St. Dev	2.738613	2.739 (It's OK to have one more SF)

Additional examples are given below for formatting Excel numbers:

X	Y	*Excel X/Y*	*Correct X/Y*
1.00	1.00	1	1.00
2.00	2.00	1	1.00

10.5. FIGURE AND GRAPH FORMAT

Data analysis and result discussion should be accompanied by figures or graphs whenever possible. Graphs can succinctly convey much information in a small space if properly done. Often the default graphs from programs like Excel are poor. Graphs must meet the following minimum criteria:

- **Type**: Choose the type of plot that best fits the experiment: linear-linear, semi-log for experiments that show exponential decay, log-log for power-law formulas, *etc*. Also, use specialized pre-printed charts when available, *i.e.*, thermodynamic P-h charts for specific refrigerants, Mollier charts for steam, *etc*.

- **Axes**: Usually, the independent variable is located on the X-axis and the dependent variable on the Y-axis. In either portrait or landscape mode, the X-axis is the horizontal axis, the Y-axis vertical.

Note: When landscape plots are included in the report, they are oriented so that the staple is in the upper-right-hand corner, *i.e.*, the top of the plot is along with the 'binding' if the report was a book. Upside down plots will not be accepted.

- **Scales:** The numeric scales and divisions chosen for the axes should span the range of the data and have logical, easy-to-read major and minor numbers. Note that most spread-sheet default initial plots (*i.e.*, Excel) generate their divisions that are usually poor. Select options to enforce divisions that best illustrate the data. Clear the default grey background – save printer toner or inkjet cartridges.

- **Labels and Title:** Every plot must have a title. The axes have to be labeled with units. If plot symbols or different line types are used, they must be identified in a legend or caption.

- **Data and Equations:** Measured data are to be shown as individual points; results from least-squares fits or theoretical equations are to be shown as continuous lines.

- **Figure Locations:** Figures must be placed next to the relevant text in the main body of the report to provide context. Number figures consecutively following their appearance in the text.

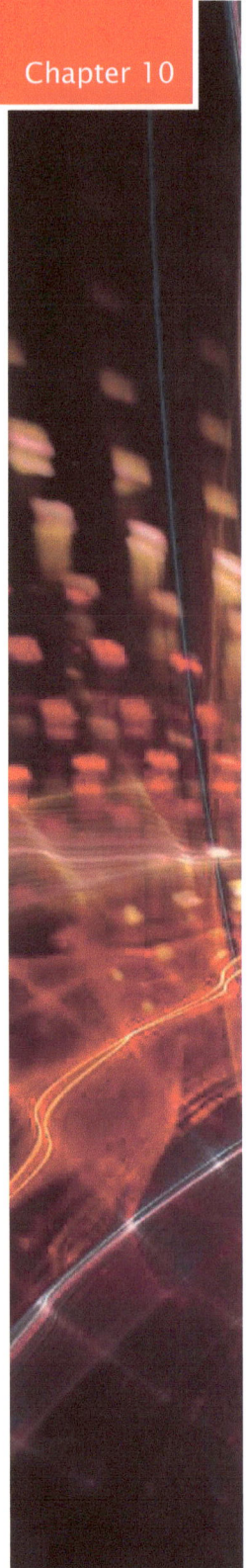

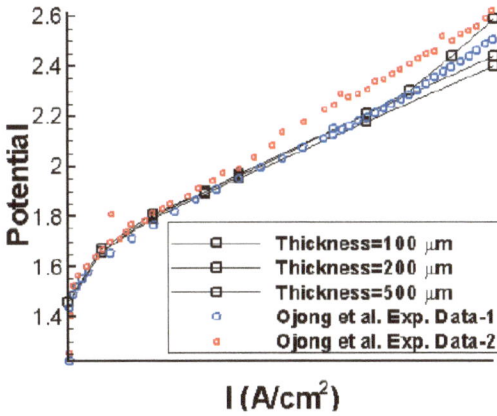

Fig. (10.1). Performance of a water electrolyzer in comparison with experimental data.

 # Example 3: List three improvements for Figure 10.1.

Solution (suggestions):

1.) *Add the unit for "Potential" in the y axis;*
2.) *Add the Ticks and Label in the x-axis;*
3.) *Use different symbols or colors to differentiate from each other for the first three curves.*

An improved figure is shown below in Figure 10.2.

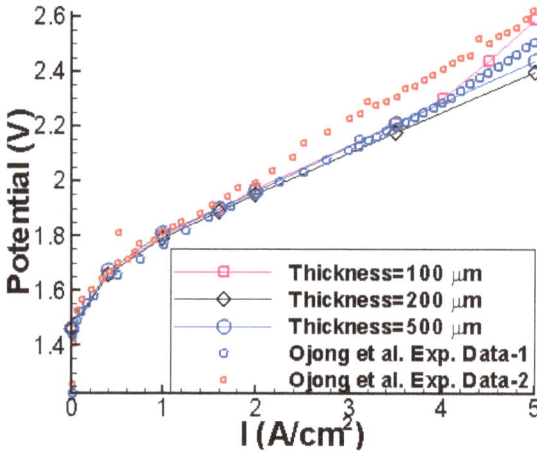

Fig. (10.2). Performance of a water electrolyzer in comparison with experimental data [3].

For **further** readings regarding technical writing, we refer the interested readers to the references in [4-8].

10.6. QUESTIONS

1. Who will be reading your lab report?

2. Which two sections are typically read first?

3. What is the purpose of an abstract or summary?

4. Why is noting the motivation for a lab report important?

5. List the major components in a well-written report;

6. List the major requirements for graphs;

7. What is one example organization that provides citation guidelines for their publications?

8. List 3-5 primary units and secondary units;

9. What is the significant figure?

10. What is the difference between accuracy and precision?

11. What is dimensional *homogeneity*?

12. Please check dimensional *homogeneity* for Equations 7.7 and 7.13.

13. Where should figures be placed in the report?

14. When the average of a number of trials is reported, can we report the answer to one more significant figure than in the individual trial?

15. Give the following Excel spreadsheet (rows 5-60 are suppressed), how would you calculate the "final balance" after a "starting balance" has accrued interest at a given "annual interest rate" compounded monthly for a given "number of months" from 1 through 60 months. Use the formula $F = P(1 + i)^n$. Indicate exactly what you would type and in which cells, as well as what Excel operations you would do and how you would do them. Only edit cells in column E and note that cells B1 and B2 are inputs.

	A	B	C	D	E
1	Starting Balance:	$5000		# of Months	Final Balance
2	Annual Interest Rate:	13%		1	
3				2	
4				3	
...				...	
61				60	

16. How would you lead a cross-functional (*i.e.* multidisciplinary) team located in multiple geographic regions?

17. How would you address a conflict between two members of a team you are leading? What would you do if that doesn't work?

REFERENCES

[1]   Mars orbit. Available: https://mars.nasa.gov/msp98/news/mco990930.html

[2]   Significant figures. Available:
       http://www.chem.tamu.edu/class/fyp/mathrev/mr-sigfg.html

[3]   Q. Chen, Y. Wang, F. Yang and H. Xu, "Two-dimensional multi-physics modeling of porous transport layer in polymer electrolyte membrane electrolyzer for water splitting". International Journal of Hydrogen Energy, vol. 45, pp.32984-32994, 2020.

[4]   R. S. Figliola and D. E. Beasley, "Theory and design for mechanical measurements," ed: IOP Publishing, 2001.

[5]   J. P. Holman, "Experimental methods for engineers", 2001.

[6]   G. J. Alred, C. T. Brusaw and W. E. Oliu, "Handbook of technical writing". Bedford/st Martins, 2006.

[7]   G. H. Mills and J. A. Walter, "Technical writing". Holt Rinehart and Winston, 2018.

[8]   M. Young, "The technical writer's handbook: writing with style and clarity". University Science Books, 2002.

Appendix I

Universal Constants

Standard Gravitational Acceleration	$g = 9.80665 \text{ m/s}^2 = 32.1742 \text{ ft/s}^2$
Speed of Light	$c = 2.998 \times 10^8 \text{ m/s}$
Stefan-Boltzmann Constant	$\sigma = 5.670 \times 10^{-8} \text{ W/m}^2.\text{K}^4$
	$= 0.1712 \times 10^{-8} \text{ Btu/h.ft}^2.\text{R}^4$
Universal Gas constant	$\bar{R} = 8314.4 \text{ J/kg mole.K}$
	$= 1.9859 \text{ Btu/lbmole.R}$
	$= 1545.35 \text{ ft.lbf/lbmole.R}$

Conversion Factor	To Convert from	To	Multiply by
Energy	Btu	J	1055.0
	cal	J	4.186
	kWh	kJ	3600
	ft.lbf	Btu	0.00128507
	hp.h	Btu	2545
Force	dyn	N	10^{-5}
	lbf	N	4.4482
Thermal Conductivity	Btu/h.ft.F	W/m.C	1.7307
Heat transfer coefficient	Btu/h.ft^2.F	W/m^2.C	5.6782
Length	ft	m	0.3048
	in	cm	2.540
	m	cm	100
	μm	m	10^{-6}
	mile	km	1.60934
Mass	lbm	kg	0.4536
	slug	lbm	32.174
	ton (metric)	kg	1000
	ton (metric)	lbm	2204.6
	ton (short)	lbm	2000
Power	Btu/h	W	0.293
	Btu/s	W	1055.04
	hp	W	745.7
	hp	ft.lbf/s	550
Pressure	atm	kPa	101.325
	bar	kPa	100
	lbf/in^2 (psi)	kPa	6.895
	atm	psi	14.696
	atm	cm Hg at 0 C	76.0
	atm	cm H$_2$O at 4 C	1033.2
Temperature	Deg. K[1]	R	9/5
	Deg. R	K	5/9
Volume	cm^3	m^3	10^{-6}
	ft^3	m^3	0.02832
	gallon (US)	m^3	0.0037854
	gallon (US)	ft^3	0.13368
	liter	m^3	10^{-3}

[1]The following relations should be used for temperature conversion:

Deg C to Deg. K	Deg K = Deg C + 273.15
Deg. F to Deg. C	Deg. C = (5/9)(Deg. F – 32)
Deg. F to Deg R	Deg. R = Deg. F + 459.67

Mathematical Basics and Relations

1. *Definitions*

(a) Dyadic product. $\left(\vec{a}\vec{c}\right)_{ij} = a_i c_j$. ($\vec{a}\vec{c}$ is a tensor.)

(b) Double dot product.

$$\vec{\sigma} : \vec{\tau} = \sum_i \sum_j \sigma_{ij} \tau_{ji} \ .$$

(c) A tensor operating on a vector from the right yields a vector.

$$\vec{a} \cdot \vec{\tau} = \sum_i \sum_j \vec{e}_i a_j \tau_{ji} \ .$$

(d) Transpose of a tensor.

$$\left(\vec{\tau}^*\right)_{ij} = \tau_{ji} \quad \text{or} \quad \vec{\tau} \cdot \vec{a} = \vec{a} \cdot \vec{\tau}^* .$$

(e) Product of two tensors.

$$\left(\vec{\tau} \cdot \vec{\sigma}\right) \cdot \vec{v} = \vec{\tau} \cdot \left(\vec{\sigma} \cdot \vec{v}\right) \quad \text{or} \quad \left(\vec{\tau} \cdot \vec{\sigma}\right)_{ij} = \sum_k \tau_{ik} \sigma_{kj} \ .$$

(f) The divergence of a tensor is a vector.

$$\nabla \cdot \vec{\tau} = \sum_i \sum_j \vec{e}_i \left(\frac{\partial \tau_{ji}}{\partial x_j} \right).$$

(g) Laplacian of a scalar.

$$\nabla^2 \Phi = \nabla \cdot \nabla \Phi = \sum_i \left(\frac{\partial^2 \Phi}{\partial x_i^2} \right)$$

(h) Gradient of a vector. $\left(\nabla \vec{v}\right)_{ij} = \partial v_j / \partial x_i$.

(i) Laplacian of a vector. $\nabla^2 \vec{v} = \nabla \cdot \nabla \vec{v} = \nabla\left(\nabla \cdot \vec{v}\right) - \nabla \times \nabla \times \vec{v}$.

2. *Algebra*

(a) $\vec{\tau} : \left(\vec{a}\vec{b}\right) = \vec{b} \cdot \left(\vec{\tau} \cdot \vec{a}\right)$.

(b) $\left(\vec{u}\vec{v}\right) : \left(\vec{w}\vec{z}\right) = \left(\vec{u}\vec{w}\right) : \left(\vec{v}\vec{z}\right) = \left(\vec{u} \cdot \vec{z}\right)\left(\vec{v} \cdot \vec{w}\right)$.

(c) $\vec{a} \cdot \left(\vec{b}\vec{c}\right) = \left(\vec{a} \cdot \vec{b}\right)\vec{c}$.

(d) $\left(\vec{a}\vec{b}\right) \cdot \vec{c} = \vec{a}\left(\vec{b} \cdot \vec{c}\right)$.

(e) $\vec{a} \times \left(\vec{b} \times \vec{c}\right) = \vec{b}\left(\vec{a} \cdot \vec{c}\right) - \vec{c}\left(\vec{a} \cdot \vec{b}\right)$.

(f) $\vec{u} \cdot \left(\vec{v} \times \vec{w}\right) = \vec{v} \cdot \left(\vec{w} \times \vec{u}\right)$.

(g) $\left(\vec{u} \times \vec{v}\right) \cdot \left(\vec{w} \times \vec{z}\right) = \left(\vec{u} \cdot \vec{w}\right)\left(\vec{v} \cdot \vec{z}\right) - \left(\vec{u} \cdot \vec{z}\right)\left(\vec{v} \cdot \vec{w}\right)$.

(h) $\vec{v} \cdot \left(\vec{\tau}^* \cdot \vec{w}\right) = \vec{w} \cdot \left(\vec{\tau} \cdot \vec{v}\right)$.

3. *Differentiation of products*

(a) $\nabla \phi \psi = \phi \nabla \psi + \psi \nabla \phi$ (a vector).

(b) $\nabla \phi \vec{v} = \phi \nabla \vec{v} + \left(\nabla \phi\right)\vec{v}$ (a tensor).

(c) $\nabla(\vec{a}\cdot\vec{c})=\vec{a}\cdot\nabla\vec{c}+\vec{c}\cdot\nabla\vec{a}+\vec{a}\times\nabla\times\vec{c}+\vec{c}\times\nabla\times\vec{a}$

$\qquad = (\nabla c)\cdot\vec{a}+(\nabla a)\cdot\vec{c}$ (a vector).

(d) $\nabla\cdot(\phi\vec{v})=\phi\nabla\cdot\vec{v}+\vec{v}\cdot\nabla\phi$ (a scalar).

(e) $\nabla\cdot(\vec{v}\times\vec{w})=\vec{w}\cdot(\nabla\times\vec{v})-\vec{v}\cdot(\nabla\times\vec{w})$ (a scalar).

(f) $\nabla\times(\phi\vec{v})=\phi\nabla\times\vec{v}+(\nabla\phi)\times\vec{v}$ (a vector).

(g) $\nabla\times(\vec{b}\times\vec{c})=\vec{b}(\nabla\cdot\vec{c})-\vec{c}(\nabla\cdot\vec{b})+\vec{c}\cdot\nabla\vec{b}-\vec{b}\cdot\nabla\vec{c}$ (a vector).

(h) $\nabla\cdot(\vec{a}b)=(\nabla\cdot\vec{a})\vec{b}+\vec{a}\cdot\nabla b$ (a vector).

(i) $\nabla\cdot(\phi\vec{\tau})=\phi\nabla\cdot\vec{\tau}+(\nabla\phi)\cdot\vec{\tau}$ (a vector).

(j) $\nabla\cdot(\vec{u}\cdot\vec{\tau})=\vec{\tau}:\nabla\vec{u}+\vec{u}\cdot\nabla\cdot\vec{\tau}^{*}$ (a scalar).

4. *Various forms of Gauss's law (divergence theorem) and Stoke's law (dS =area
element, dl = line element, dv =volume element. Integration over a closed surface
or a closed curve is denoted by a circle through the integral sign. In the first case,*
$\overrightarrow{dS}$ *is normally outward from the surface; in the second case,* $\overrightarrow{dl}$ *and* $\overrightarrow{dS}$ *are
related by a right-hand screw rule, that is, a right-hand screw turned in the direction
of* $\overrightarrow{dl}$ *advances in the direction of* $\overrightarrow{dS}$ *.)*

(a) $\oint\overrightarrow{dS}\cdot\vec{F}=\int dv\nabla\cdot\vec{F}$.

(b) $\oint\overrightarrow{dS}\phi=\int dv\nabla\phi$.

(c) $\oint(\overrightarrow{dS}\cdot\vec{G})\vec{F}=\int dv\vec{F}\nabla\cdot\vec{G}+\int dv\vec{G}\cdot\nabla\vec{F}$.

(d) $\oint\overrightarrow{dS}\times\vec{F}=\int dv\nabla\times\vec{F}$.

(e) $\oint\overrightarrow{dS}\cdot\vec{\tau}=\int dv\nabla\cdot\vec{\tau}$.

(f) $\oint\overrightarrow{dS}\cdot(\Psi\nabla\phi-\phi\nabla\Psi)=\int dv(\Psi\nabla^{2}\phi-\phi\nabla^{2}\Psi)$.

(g) $\oint\overrightarrow{dl}\cdot\vec{F}=\int\overrightarrow{dS}\cdot\nabla\times\vec{F}$.

(h) $\oint\overrightarrow{dl}\phi=\int\overrightarrow{dS}\times\nabla\phi$.

5. *Miscellaneous*

(a) $\nabla\cdot\nabla\times\vec{E}=0$.

(b) $\nabla\times\nabla\phi=0$.

(c) $\vec{w}\cdot\nabla\vec{v}=\sum_{i}\sum_{j}\vec{e}_{i}w_{j}\dfrac{\partial v_{i}}{\partial x_{j}}$.

(d) $D/Dt=\partial/\partial t+\vec{v}\cdot\nabla$.

(e) $D\vec{v}/Dt=\partial\vec{v}/\partial t+\dfrac{1}{2}\nabla v^{2}-\vec{v}\times\nabla\times\vec{v}$ $\bigg\}$ where $\vec{v}$ is the mass-average velocity.

6. *Scalar product of vectors*

$$\vec{a} \cdot \vec{b} = \left(\sum_i a_i \vec{e}_i \right) \cdot \left(\sum_j b_j \vec{e}_j \right) = \sum_i \sum_j a_i b_j \left(\vec{e}_i \cdot \vec{e}_j \right) = \sum_i \sum_j a_i b_j \delta_{ij} = \sum_i a_i b_i .$$

7. *Vector product of vectors*

$$\vec{a} \times \vec{b} = \begin{vmatrix} \vec{e}_1 & \vec{e}_2 & \vec{e}_3 \\ a_1 & a_2 & a_3 \\ b_1 & b_2 & b_3 \end{vmatrix} .$$

8. *Multiple products*

(a) $\vec{a} \cdot \left(\vec{b} \times \vec{c} \right) = \vec{b} \cdot \left(\vec{c} \times \vec{a} \right) = \vec{c} \cdot \left(\vec{a} \times \vec{b} \right).$

(b) $\vec{a} \times \left(\vec{b} \times \vec{c} \right) = \left(\vec{a} \cdot \vec{c} \right) \vec{b} - \left(\vec{a} \cdot \vec{b} \right) \vec{c} .$

(c) $\left(\vec{a} \times \vec{b} \right) \times \vec{c} = \left(\vec{a} \cdot \vec{c} \right) \vec{b} - \left(\vec{b} \cdot \vec{c} \right) \vec{a} .$

(d) $\left(\vec{a} \times \vec{b} \right) \cdot \left(\vec{c} \times \vec{d} \right) = \left(\vec{a} \cdot \vec{c} \right)\left(\vec{b} \cdot \vec{d} \right) - \left(\vec{a} \cdot \vec{d} \right)\left(\vec{b} \cdot \vec{c} \right).$

9. *Identity tensor*

$$\vec{\delta} = \sum_i \sum_j \delta_{ij} \vec{e}_i \vec{e}_j = \sum_i \vec{e}_i \vec{e}_i = \begin{bmatrix} 1 & 0 & 0 \\ 0 & 1 & 0 \\ 0 & 0 & 1 \end{bmatrix} .$$

10. *Vector-differential operators*

(a) Gradient.

$$\nabla = \vec{e}_1 \frac{\partial}{\partial x_1} + \vec{e}_2 \frac{\partial}{\partial x_2} + \vec{e}_3 \frac{\partial}{\partial x_3} = \sum_i \vec{e}_i \frac{\partial}{\partial x_i} .$$

(b) Divergence.

$$\nabla \cdot \vec{v} = \frac{\partial v_1}{\partial x_1} + \frac{\partial v_2}{\partial x_2} + \frac{\partial v_3}{\partial x_3} = \sum_i \frac{\partial v_i}{\partial x_i} .$$

(c) Curl.

$$\nabla \times \vec{v} = \left(\frac{\partial v_3}{\partial x_2} - \frac{\partial v_2}{\partial x_3} \right) \vec{e}_1 + \left(\frac{\partial v_1}{\partial x_3} - \frac{\partial v_3}{\partial x_1} \right) \vec{e}_2 + \left(\frac{\partial v_2}{\partial x_1} - \frac{\partial v_1}{\partial x_2} \right) \vec{e}_3 .$$

(d) Laplacian.

$$\nabla \cdot \nabla = \nabla^2 = \frac{\partial^2}{\partial x_1^2} + \frac{\partial^2}{\partial x_2^2} + \frac{\partial^2}{\partial x_3^2} = \sum_i \frac{\partial^2}{\partial x_i^2}$$

(e)   Material derivative.

$$\frac{D}{Dt} = \frac{\partial}{\partial t} + \vec{v} \cdot \nabla = \frac{\partial}{\partial t} + \sum_i v_i \frac{\partial}{\partial x_i}$$

*Sources:

1.) Y. A. Cengel, M. A. Boles, Thermodynamics, An engineering approach, 6th Ed. (McGraw-Hill, 2007).

2.) NIST Chemistry WebBook.

3.) Wang, Y., & Chen, K. S. (2013). PEM fuel cells: thermal and water management fundamentals. Momentum Press.

4.) J. Newman and K. E. Thomas-Alyea, Electrochemical Systems, Wiley-Interscience; 3 edition (May 27, 2004).

APPENDIX II. THERMODYNAMIC PROPERTIES OF AIR, HYDROGEN GAS, AND WATER VAPOR

Table A.I-1. Ideal-gas properties of air (k: 1.406 ~ 1.392)

T	H†	U†	S**	$C_p{}^*$	$C_v{}^*$
[K]	[kJ/kg]	[kJ/kg]	[kJ/(kg·K)]	[kJ/(kg·K)]	[kJ/(kg·K)]
230	232.0409	165.5811	1.776346	0.9939	0.7069
240	242.1921	172.9011	1.789524	0.9952	0.7082
243	245.2402	175.0744	1.793309	0.9956	0.7086
250	251.7154	180.1442	1.802048	0.9966	0.7096
260	262.4987	187.4031	1.814533	0.9980	0.7110
270	272.6462	194.6583	1.826454	0.9994	0.7124
273	275.6964	196.8337	1.830082	0.9998	0.7128
280	282.8037	201.9056	1.838426	1.0008	0.7138
290	292.9554	209.1629	1.849896	1.0023	0.7153
300	303.1092	216.4244	1.861485	1.0038	0.7168
310	313.2751	223.688	1.872625	1.0053	0.7183
320	323.4352	230.9558	1.883931	1.0068	0.7198
330	333.6074	238.2278	1.894828	1.0083	0.7213
340	343.7838	245.5098	1.905932	1.0099	0.7229
350	353.9644	252.7902	1.916853	1.0115	0.7245
353	357.023	254.9819	1.920122	1.0120	0.7250
360	364.1492	260.0906	1.927815	1.0131	0.7261
370	382.2482	267.3873	1.938616	1.0147	0.7277
380	384.5593	274.6982	1.949677	1.0163	0.7293
390	394.7746	282.0212	1.960601	1.0180	0.7310
393	397.8395	284.2192	1.963892	1.0185	0.7315
400	404.9983	289.3584	1.971593	1.0197	0.7327

Sources:

*Based on a third-degree polynomial equation in Y. A. Cengel, M. A. Boles, Thermodynamics, An engineering approach, 6th Ed. (McGraw-Hill, 2007).

† Based on data for $O_2(21\%) + N_2(79\%)$ from NIST Chemistry WebBook.

** Based on $S = \sum_k n_k \left(c_{vk} \log T + R \log \dfrac{V}{n_k} \right)$ using data from NIST Chemistry WebBook.

Table A.I-2. Ideal-gas properties of Hydrogen gas (k: 1.404 ~ 1.403)

T	H	U	S	$C_p{}^{†}$	$C_v{}^{†}$
[K]	[kJ/kg]	[kJ/kg]	[kJ/(kg·K)]	[kJ/(kg·K)]	[kJ/(kg·K)]
260	3655.9716	2583.9832	62.8192	14.1329	10.0089
270	3798.3412	2684.6836	63.3564	14.1883	10.0643
273†	3842.3887	2715.7560	63.5690	14.2033	10.0793
280	3941.2068	2786.3762	63.8753	14.2357	10.1117
290	4084.0725	2888.0687	64.3764	14.2763	10.1523
300	4227.4342	2989.7613	64.8620	14.3111	10.1871
310	4370.7959	3092.1979	65.3251	14.3410	10.2170
320	4514.1576	3194.6346	65.7881	14.3666	10.2426
330	4658.0154	3297.0713	66.2239	14.3886	10.2646
340	4801.8731	3399.5079	66.6597	14.4074	10.2834
350	4946.2270	3502.4406	67.0717	14.4236	10.2996
353†	4989.4351	3532.7869	67.2526	14.4280	10.3040
360	5090.5808	3605.3733	67.4837	14.4375	10.3135
370	5234.6866	3708.5541	67.8738	14.4494	10.3254
380	5378.7924	3811.7348	68.2640	14.4596	10.3356
390	5523.3942	3914.9156	68.6345	14.4683	10.3443
393†	5567.4961	3945.7895	68.8038	14.4707	10.3467
400	5667.9961	4018.0963	69.0051	14.4759	10.3519

Table A.I-3. Ideal-gas properties of Oxygen (k: 1.409 ~ 1.375)

T	H	U	S	$C_p^{\dagger}$	$C_v^{\dagger}$
[K]	[kJ/kg]	[kJ/kg]	[kJ/(kg·K)]	[kJ/(kg·K)]	[kJ/(kg·K)]
230	209.1953	149.4431	6.1709	0.9108	0.6510
240	218.2582	155.9121	6.2095	0.9115	0.6517
243[†]	220.8715	157.8291	6.2240	0.9117	0.6519
250	227.3523	162.4123	6.2466	0.9123	0.6525
260	236.4464	168.9126	6.2823	0.9132	0.6534
270	245.5717	175.4128	6.3167	0.9143	0.6545
273[†]	248.2640	177.6587	6.3303	0.9147	0.6549
280	254.6971	181.9443	6.3500	0.9156	0.6558
290	263.8536	188.5071	6.3821	0.9170	0.6572
300	273.0102	195.0698	6.4131	0.9185	0.6587
310	282.1981	201.6638	6.4433	0.9202	0.6604
320	291.4172	208.2578	6.4725	0.9220	0.6622
330	300.6363	214.9143	6.5009	0.9240	0.6642
340	309.8866	221.5708	6.5285	0.9261	0.6663
350	319.1682	228.2273	6.5554	0.9283	0.6685
353[†]	321.9528	230.2516	6.5670	0.9289	0.6691
360	328.4811	234.9463	6.5816	0.9306	0.6708
370	337.7939	241.6653	6.6072	0.9330	0.6732
380	347.1693	248.4156	6.6322	0.9355	0.6757
390	356.5446	255.1971	6.6566	0.9381	0.6783
393[†]	359.3043	257.1991	6.6672	0.9389	0.6791
400	365.9825	262.0098	6.6804	0.9408	0.6810

*Sources:

[1] Y. A. Cengel, M. A. Boles, "Thermodynamics, An engineering approach", 6th Ed. (McGraw-Hill, 2007).

[2] NIST Chemistry WebBook.

[3] Y. Wang and K. S. Chen, "PEM fuel cells: thermal and water management fundamentals". Momentum Press, 2013.

Table A3.1 Mass conservation equations in the cartesian, cylindrical and spherical coordinates.

Cartesian coordinates (x, y, z):

$$\frac{\partial \rho}{\partial t} + \frac{\partial}{\partial x}(\rho u_x) + \frac{\partial}{\partial y}(\rho u_y) + \frac{\partial}{\partial z}(\rho u_z) = 0$$

Cylindrical coordinates (r, θ, z):

$$\frac{\partial \rho}{\partial t} + \frac{1}{r}\frac{\partial}{\partial r}(\rho r u_r) + \frac{1}{r}\frac{\partial}{\partial \theta}(\rho u_\theta) + \frac{\partial}{\partial z}(\rho u_z) = 0$$

Spherical coordinates (r, θ, ϕ):

$$\frac{\partial \rho}{\partial t} + \frac{1}{r^2}\frac{\partial}{\partial r}(\rho r^2 u_r) + \frac{1}{r\sin\theta}\frac{\partial}{\partial \theta}(\rho u_\theta \sin\theta) + \frac{1}{r\sin\theta}\frac{\partial}{\partial \phi}(\rho u_\phi) = 0$$

Table A3.2 Momentum equations in the cartesian coordinate.

Cartesian coordinates (x, y, z) x-direction:

$$\rho\left(\frac{\partial u_x}{\partial t}+u_x\frac{\partial u_x}{\partial x}+u_y\frac{\partial u_x}{\partial y}+u_z\frac{\partial u_x}{\partial z}\right)=-\frac{\partial p}{\partial x}+\mu\left(\frac{\partial^2 u_x}{\partial x^2}+\frac{\partial^2 u_x}{\partial y^2}+\frac{\partial^2 u_x}{\partial z^2}\right)+\rho g_x$$

Cartesian coordinates (x, y, z) y-direction:

$$\rho\left(\frac{\partial u_y}{\partial t}+u_x\frac{\partial u_y}{\partial x}+u_y\frac{\partial u_y}{\partial y}+u_z\frac{\partial u_y}{\partial z}\right)=-\frac{\partial p}{\partial y}+\mu\left(\frac{\partial^2 u_y}{\partial x^2}+\frac{\partial^2 u_y}{\partial y^2}+\frac{\partial^2 u_y}{\partial z^2}\right)+\rho g_y$$

Cartesian coordinates (x, y, z) z-direction:

$$\rho\left(\frac{\partial u_z}{\partial t}+u_x\frac{\partial u_z}{\partial x}+u_y\frac{\partial u_z}{\partial y}+u_z\frac{\partial u_z}{\partial z}\right)=-\frac{\partial p}{\partial z}+\mu\left(\frac{\partial^2 u_z}{\partial x^2}+\frac{\partial^2 u_z}{\partial y^2}+\frac{\partial^2 u_z}{\partial z^2}\right)+\rho g_z$$

Table A3.3 Momentum equations in the cylindrical coordinate.

Cylindrical coordinates (r, θ, z) r-direction:

$$\rho\left(\frac{\partial u_r}{\partial t}+u_r\frac{\partial u_r}{\partial r}+\frac{u_\theta}{r}\frac{\partial u_r}{\partial \theta}-\frac{u_\theta^2}{r}+u_z\frac{\partial u_r}{\partial z}\right)$$
$$=-\frac{\partial p}{\partial r}+\mu\left[\frac{\partial}{\partial r}\left(\frac{1}{r}\frac{\partial}{\partial r}\left(ru_r\right)\right)+\frac{1}{r^2}\frac{\partial^2 u_r}{\partial \theta^2}-\frac{2}{r^2}\frac{\partial u_\theta}{\partial \theta}+\frac{\partial^2 u_r}{\partial z^2}\right]+\rho g_r$$

Cylindrical coordinates (r, θ, z) θ -direction:

$$\rho\left(\frac{\partial u_\theta}{\partial t}+u_r\frac{\partial u_\theta}{\partial r}+\frac{u_\theta}{r}\frac{\partial u_\theta}{\partial \theta}-\frac{u_r u_\theta}{r}+u_z\frac{\partial u_\theta}{\partial z}\right)$$
$$=-\frac{1}{r}\frac{\partial p}{\partial \theta}+\mu\left[\frac{\partial}{\partial r}\left(\frac{1}{r}\frac{\partial}{\partial r}\left(ru_\theta\right)\right)+\frac{1}{r^2}\frac{\partial^2 u_\theta}{\partial \theta^2}+\frac{2}{r^2}\frac{\partial u_r}{\partial \theta}+\frac{\partial^2 u_\theta}{\partial z^2}\right]+\rho g_\theta$$

Cylindrical coordinates (r, θ, z) z-direction:

$$\rho\left(\frac{\partial u_z}{\partial t}+u_r\frac{\partial u_z}{\partial r}+\frac{u_\theta}{r}\frac{\partial u_z}{\partial \theta}+u_z\frac{\partial u_z}{\partial z}\right)$$

$$=-\frac{\partial p}{\partial z}+\mu\left[\frac{1}{r}\frac{\partial}{\partial r}\left(r\frac{\partial u_z}{\partial r}\right)+\frac{1}{r^2}\frac{\partial^2 u_z}{\partial \theta^2}+\frac{\partial^2 u_z}{\partial z^2}\right]+\rho g_z$$

Table A3.4 Momentum equations in the spherical coordinate.

Spherical coordinates (r, θ, ϕ) r-direction:

$$\rho\left(\frac{\partial u_r}{\partial t}+u_r\frac{\partial u_r}{\partial r}+\frac{u_\theta}{r}\frac{\partial u_r}{\partial \theta}+\frac{u_\phi}{r\sin\theta}\frac{\partial u_r}{\partial \phi}-\frac{u_\theta^2+u_\phi^2}{r}\right)$$

$$=-\frac{\partial p}{\partial r}+\mu\left[\nabla^2 u_r-\frac{2}{r^2}u_r-\frac{2}{r^2}\frac{\partial u_\theta}{\partial \theta}-\frac{2}{r^2}u_\theta\cot\theta-\frac{2}{r^2\sin\theta}\frac{\partial u_\phi}{\partial \phi}\right]+\rho g_r$$

Spherical coordinates (r, θ, ϕ) θ-direction:

$$\rho\left(\frac{\partial u_\theta}{\partial t}+u_r\frac{\partial u_\theta}{\partial r}+\frac{u_\theta}{r}\frac{\partial u_\theta}{\partial \theta}+\frac{u_\phi}{r\sin\theta}\frac{\partial u_\theta}{\partial \phi}+\frac{u_r u_\theta}{r}-\frac{u_\phi^2\cot\theta}{r}\right)$$

$$=-\frac{1}{r}\frac{\partial p}{\partial \theta}+\mu\left[\nabla^2 u_\theta+\frac{2}{r^2}\frac{\partial u_r}{\partial \theta}-\frac{u_\theta}{r^2\sin^2\theta}-\frac{2\cos\theta}{r^2\sin^2\theta}\frac{\partial u_\phi}{\partial \phi}\right]+\rho g_\theta$$

Spherical coordinates (r, θ, ϕ) ϕ-direction:

$$\rho\left(\frac{\partial u_\phi}{\partial t}+u_r\frac{\partial u_\phi}{\partial r}+\frac{u_\theta}{r}\frac{\partial u_\phi}{\partial \theta}+\frac{u_\phi}{r\sin\theta}\frac{\partial u_\phi}{\partial \phi}+\frac{u_\phi u_r}{r}+\frac{u_\theta u_\phi}{r}\cot\theta\right)$$

$$=-\frac{1}{r\sin\theta}\frac{\partial p}{\partial \phi}+\mu\left[\nabla^2 u_\phi-\frac{u_\phi}{r^2\sin^2\theta}+\frac{2}{r^2\sin\theta}\frac{\partial u_r}{\partial \phi}+\frac{2\cos\theta}{r^2\sin^2\theta}\frac{\partial u_\theta}{\partial \phi}\right]+\rho g_\phi$$

Table A3.5 Energy equations in the cartesian, cylindrical and spherical coordinates

Cartesian coordinates (x, y, z):

$$\rho c_p \left(\frac{\partial T}{\partial t} + u_x \frac{\partial T}{\partial x} + u_y \frac{\partial T}{\partial y} + u_z \frac{\partial T}{\partial z} \right) = k \left(\frac{\partial^2 T}{\partial x^2} + \frac{\partial^2 T}{\partial y^2} + \frac{\partial^2 T}{\partial z^2} \right)$$

$$-\beta T \left(\frac{\partial p}{\partial t} + u_x \frac{\partial p}{\partial x} + u_y \frac{\partial p}{\partial y} + u_z \frac{\partial p}{\partial z} \right) + \mu \Phi + \dot{q}$$

$$\Phi = 2 \left[\left(\frac{\partial u_x}{\partial x} \right)^2 + \left(\frac{\partial u_y}{\partial y} \right)^2 + \left(\frac{\partial u_z}{\partial z} \right)^2 \right] + \left[\left(\frac{\partial u_x}{\partial y} + \frac{\partial u_y}{\partial x} \right)^2 + \left(\frac{\partial u_y}{\partial z} + \frac{\partial u_z}{\partial y} \right)^2 + \left(\frac{\partial u_z}{\partial x} + \frac{\partial u_x}{\partial z} \right)^2 \right]$$

$$- \frac{2}{3} \left(\frac{\partial u_x}{\partial x} + \frac{\partial u_y}{\partial y} + \frac{\partial u_z}{\partial z} \right)^2$$

Cylindrical coordinates (r, θ, z):

$$\rho c_p \left(\frac{\partial T}{\partial t} + u_r \frac{\partial T}{\partial r} + \frac{u_\theta}{r} \frac{\partial T}{\partial \theta} + u_z \frac{\partial T}{\partial z} \right) = \frac{k}{r} \frac{\partial}{\partial r} \left(r \frac{\partial T}{\partial r} \right) + \frac{k}{r^2} \frac{\partial^2 T}{\partial \theta^2} + k \frac{\partial^2 T}{\partial z^2}$$

$$-\beta T \left(\frac{\partial p}{\partial t} + u_r \frac{\partial p}{\partial r} + \frac{u_\theta}{r} \frac{\partial p}{\partial \theta} + u_z \frac{\partial p}{\partial z} \right) + \mu \Phi + \dot{q}$$

$$\Phi = 2 \left[\left(\frac{\partial u_r}{\partial r} \right)^2 + \left(\frac{1}{r} \frac{\partial u_\theta}{\partial \theta} + \frac{u_r}{r} \right)^2 + \left(\frac{\partial u_z}{\partial z} \right)^2 \right] + \left[r \frac{\partial}{\partial r} \left(\frac{u_\theta}{r} \right) + \frac{1}{r} \left(\frac{\partial u_r}{\partial \theta} \right) \right]^2$$

$$+ \left[\frac{1}{r} \frac{\partial u_z}{\partial \theta} + \frac{\partial u_x}{\partial z} \right]^2 + \left[\frac{\partial u_r}{\partial z} + \frac{\partial u_z}{\partial r} \right]^2 - \frac{2}{3} \left[\frac{1}{r} \frac{\partial}{\partial r} (r u_r) + \frac{1}{r} \frac{\partial u_\theta}{\partial \theta} + \frac{\partial u_z}{\partial z} \right]^2$$

Spherical coordinates (r, θ, ϕ):

$$\rho c_p \left(\frac{\partial T}{\partial t} + u_r \frac{\partial T}{\partial r} + \frac{u_\theta}{r} \frac{\partial T}{\partial \theta} + \frac{u_\phi}{r \sin \theta} \frac{\partial T}{\partial \phi} \right) = \frac{k}{r^2} \frac{\partial}{\partial r} \left(r^2 \frac{\partial T}{\partial r} \right) + \frac{k}{r^2 \sin \theta} \frac{\partial}{\partial \theta} \left(\sin \theta \frac{\partial T}{\partial \theta} \right) + \frac{k}{r^2 \sin^2 \theta} \frac{\partial^2 T}{\partial \phi^2}$$

$$-\beta T \left(\frac{\partial p}{\partial t} + u_r \frac{\partial p}{\partial r} + \frac{u_\theta}{r} \frac{\partial p}{\partial \theta} + \frac{u_\phi}{r \sin \theta} \frac{\partial p}{\partial \phi} \right) + \mu \Phi + \dot{q}$$

$$\Phi = 2 \left[\left(\frac{\partial u_r}{\partial r} \right)^2 + \left(\frac{1}{r} \frac{\partial u_\theta}{\partial \theta} + \frac{u_r}{r} \right)^2 + \left(\frac{1}{r \sin \theta} \frac{\partial u_\phi}{\partial \phi} + \frac{u_r}{r} + \frac{u_\theta \cot \theta}{r} \right)^2 \right] + \left[r \frac{\partial}{\partial r} \left(\frac{u_\theta}{r} \right) + \frac{1}{r} \frac{\partial u_r}{\partial \theta} \right]^2$$

$$+ \left[\frac{\sin \theta}{r} \frac{\partial}{\partial \theta} \left(\frac{u_\phi}{\sin \theta} \right) + \frac{1}{r \sin \theta} \frac{\partial u_\theta}{\partial \phi} \right]^2 + \left[\frac{1}{r \sin \theta} \frac{\partial u_r}{\partial \phi} + r \frac{\partial}{\partial \theta} \left(\frac{u_\phi}{r} \right) \right]^2$$

$$- \frac{2}{3} \left[\frac{1}{r^2} \frac{\partial}{\partial r} (r^2 u_r) + \frac{1}{r \sin \theta} \frac{\partial}{\partial \theta} (u_\theta \sin \theta) + \frac{1}{r \sin \theta} \frac{\partial u_\phi}{\partial \phi} \right]^2$$

Table A3.6 Species equations in the cartesian, cylindrical and spherical coordinates

---

Cartesian coordinates (x, y, z):

$$\frac{\partial C_i}{\partial t} + u_x \frac{\partial C_i}{\partial x} + u_y \frac{\partial C_i}{\partial y} + u_z \frac{\partial C_i}{\partial z} = D_i \left[\frac{\partial^2 C_i}{\partial x^2} + \frac{\partial^2 C_i}{\partial y^2} + \frac{\partial^2 C_i}{\partial z^2} \right] + R_{Vi}$$

Cylindrical coordinates (r, θ, z):

$$\frac{\partial C_i}{\partial t} + u_r \frac{\partial C_i}{\partial r} + \frac{u_\theta}{r} \frac{\partial C_i}{\partial \theta} + u_z \frac{\partial C_i}{\partial z} = D_i \left[\frac{1}{r}\frac{\partial}{\partial r}\left(r \frac{\partial C_i}{\partial r} \right) + \frac{1}{r^2}\frac{\partial^2 C_i}{\partial \theta^2} + \frac{\partial^2 C_i}{\partial z^2} \right] + R_{Vi}$$

Spherical coordinates (r, θ, ϕ):

$$\frac{\partial C_i}{\partial t} + u_r \frac{\partial C_i}{\partial r} + \frac{u_\theta}{r} \frac{\partial C_i}{\partial \theta} + \frac{u_\phi}{r\sin\theta} \frac{\partial C_i}{\partial \phi}$$

$$= D_i \left[\frac{1}{r^2}\frac{\partial}{\partial r}\left(r^2 \frac{\partial C_i}{\partial r} \right) + \frac{1}{r^2 \sin\theta}\frac{\partial}{\partial \theta}\left(\sin\theta \frac{\partial C_i}{\partial \theta} \right) + \frac{1}{r^2 \sin^2\theta}\frac{\partial^2 C_i}{\partial \phi^2} \right] + R_{Vi}$$

---

*Sources:

[1] Y. Wang and K. S. Chen, "PEM fuel cells: thermal and water management fundamentals". Momentum Press, 2013.

T	ρ	$\mu \cdot 10^7$	$v \cdot 10^6$	$k \cdot 10^3$	$\alpha \cdot 10^6$	
(K)	(kg/m^3)	$(N \cdot s/m^2)$	(m^2/s)	$(W/m \cdot K)$	(m^2/s)	Pr
Air						
100	3.5562	71.1	2.00	9.34	2.54	0.786
150	2.3364	103.4	4.426	13.8	5.84	0.758
200	1.7458	132.5	7.590	18.1	10.3	0.737
250	1.3947	159.6	11.44	22.3	15.9	0.720
300	1.1614	184.6	15.89	26.3	22.5	0.707
350	0.9950	208.2	20.92	30.0	29.9	0.700
400	0.8711	230.1	26.41	33.8	38.3	0.690
450	0.7740	250.7	32.39	37.3	47.2	0.686
500	0.6964	270.1	38.79	40.7	56.7	0.684
550	0.6329	288.4	45.57	43.9	66.7	0.683
600	0.5804	305.8	52.69	46.9	76.9	0.685
650	0.5356	322.5	60.21	49.7	87.3	0.690
700	0.4975	338.8	68.10	52.4	98.0	0.695
750	0.4643	354.6	76.37	54.9	109	0.702
800	0.4354	369.8	84.93	57.3	120	0.709
850	0.4097	384.3	93.80	59.6	131	0.716
900	0.3868	398.1	102.9	62.0	143	0.720
950	0.3666	411.3	112.2	64.3	155	0.723
1000	0.3482	424.4	121.9	66.7	168	0.726
1100	0.3166	449.0	141.8	71.5	195	0.728
1200	0.2902	473.0	162.9	76.3	224	0.728
1300	0.2679	496.0	185.1	82	238	0.719

1400	0.2488	530	213	91	303	0.703
1500	0.2322	557	240	100	350	0.685
1600	0.2177	584	268	106	390	0.688
1700	0.2049	611	298	113	435	0.685
1800	0.1935	637	329	120	482	0.683
1900	0.1833	663	362	128	534	0.677
2000	0.1741	689	396	137	589	0.672
2100	0.1658	715	431	147	646	0.667
2200	0.1582	740	468	160	714	0.655
2300	0.1513	766	506	175	783	0.647
2400	0.1448	792	547	196	869	0.630
2500	0.1389	818	589	222	960	0.613
3000	0.1135	955	841	486	1570	0.536

T	ρ	$\mu \cdot 10^7$	$v \cdot 10^6$	$k \cdot 10^3$	$\alpha \cdot 10^6$	
(K)	(kg/m^3)	$(N \cdot s/m^2)$	(m^2/s)	$(W/m \cdot K)$	(m^2/s)	*Pr*
Carbon Dioxide (CO_2)						
280	1.9022	140	7.36	15.20	9.63	0.765
300	1.7730	149	8.40	16.55	11.0	0.766
320	1.6609	156	9.39	18.05	12.5	0.754
340	1.5618	165	10.6	19.70	14.2	0.746
360	1.4743	173	11.7	21.2	15.8	0.741
380	1.3961	181	13.0	22.75	17.6	0.737
400	1.3257	190	14.3	24.3	19.5	0.737
450	1.1782	210	17.8	28.3	24.5	0.728
500	1.0594	231	21.8	32.5	30.1	0.725
550	0.9625	251	26.1	36.6	36.2	0.721

600	0.8826	270	30.6	40.7	42.7	0.717
650	0.8143	288	35.4	44.5	49.7	0.712
700	0.7564	305	40.3	48.1	56.3	0.717
750	0.7057	321	45.5	51.7	63.7	0.714
800	0.6614	337	51.0	55.1	71.2	0.716

Carbon Monoxide (CO)

200	1.6888	127	7.52	17.0	9.63	0.781
220	1.5341	137	8.93	19.0	11.9	0.753
240	1.4055	147	10.5	20.6	14.1	0.744
260	1.2967	157	12.1	22.1	16.3	0.741
280	1.2038	166	13.8	23.6	18.8	0.733
300	1.1233	175	15.6	25.0	21.3	0.730
320	1.0529	184	17.5	26.3	23.9	0.730
340	0.9909	193	19.5	27.8	26.9	0.725
360	0.9357	202	21.6	29.1	29.8	0.725
380	0.8864	210	23.7	30.5	32.9	0.729
400	0.8421	218	25.9	31.8	36.0	0.719
450	0.7483	237	31.7	35.0	44.3	0.714
500	0.67352	254	37.7	38.1	53.1	0.710
550	0.61226	271	44.3	41.1	62.4	0.710
600	0.56126	286	51.0	44.0	72.1	0.707
650	0.51806	301	58.1	47.0	82.4	0.705
700	0.48102	315	65.5	50.0	93.3	0.702
750	0.44899	329	73.3	52.8	104	0.702
800	0.42095	343	81.5	55.5	116	0.705

T	ρ	$\mu \cdot 10^7$	$v \cdot 10^6$	$k \cdot 10^3$	$\alpha \cdot 10^6$	
(K)	(kg/m³)	(N·s/m²)	(m²/s)	(W/m·K)	(m²/s)	*Pr*
Hydrogen (H₂)						
100	0.24255	42.1	17.4	67.0	24.6	0.707
150	0.16156	56.0	34.7	101	49.6	0.699
200	0.12115	68.1	56.2	131	79.9	0.704
250	0.09693	78.9	81.4	157	115	0.707
300	0.08078	89.6	111	183	158	0.701
350	0.06924	98.8	143	204	204	0.700
400	0.06059	108.2	179	226	258	0.695
450	0.05386	117.2	218	247	316	0.689
500	0.04848	126.4	261	266	378	0.691
550	0.04407	134.3	305	285	445	0.685
600	0.04040	142.4	352	305	519	0.678
700	0.03463	157.8	456	342	676	0.675
800	0.03030	172.4	569	378	849	0.670
900	0.02694	186.5	692	412	1030	0.671
1000	0.02424	201.3	830	448	1230	0.673
1100	0.02204	213.0	966	488	1460	0.662
1200	0.02020	226.2	1120	528	1700	0.659
1300	0.01865	238.5	1279	568	1955	0.655
1400	0.01732	250.7	1447	610	2230	0.650
1500	0.01616	262.7	1626	655	2530	0.643
1600	0.0152	273.7	1801	697	2815	0.639
1700	0.0143	284.9	1992	742	3130	0.637
1800	0.0135	296.1	2193	786	3435	0.639

| 1900 | 0.0128 | 307.2 | 2400 | 835 | 3730 | 0.643 |
| 2000 | 0.0121 | 318.2 | 2630 | 878 | 3975 | 0.661 |

T	ρ	$\mu \cdot 10^7$	$v \cdot 10^6$	$k \cdot 10^3$	$\alpha \cdot 10^6$	
(K)	(kg/m^3)	$(N \cdot s/m^2)$	(m^2/s)	$(W/m \cdot K)$	(m^2/s)	Pr
Nitrogen (N$_2$)						
100	3.4388	68.8	2.00	9.58	2.60	0.768
150	2.2594	100.6	4.45	13.9	5.86	0.759
200	1.6883	129.2	7.65	18.3	10.4	0.736
250	1.3488	154.9	11.48	22.2	15.8	0.727
300	1.1233	178.2	15.86	25.9	22.1	0.716
350	0.9625	200.0	20.78	29.3	29.2	0.711
400	0.8425	220.4	26.16	32.7	37.1	0.704
450	0.7485	239.6	32.01	35.8	45.6	0.703
500	0.6739	257.7	38.24	38.9	54.7	0.700
550	0.6124	274.7	44.86	41.7	63.9	0.702
600	0.5615	290.8	51.79	44.6	73.9	0.701
700	0.4812	321.0	66.71	49.9	94.4	0.706
800	0.4211	349.1	82.90	54.8	116	0.715
900	0.3743	375.3	100.3	59.7	139	0.721
1000	0.3368	399.9	118.7	64.7	165	0.721
1100	0.3062	423.2	138.2	70.0	193	0.718
1200	0.2807	445.3	158.6	75.8	224	0.707
1300	0.2591	466.2	179.9	81.0	256	0.701

T	ρ	$\mu \cdot 10^7$	$v \cdot 10^6$	$k \cdot 10^3$	$\alpha \cdot 10^6$	
(K)	(kg/m³)	(N·s/m²)	(m²/s)	(W/m·K)	(m²/s)	Pr
Oxygen (O₂)						
100	3.945	76.4	1.94	9.25	2.44	0.796
150	2.585	114.8	4.44	13.8	5.80	0.766
200	1.930	147.5	7.64	18.3	10.4	0.737
250	1.542	178.6	11.58	22.6	16.0	0.723
300	1.284	207.2	16.14	26.8	22.7	0.711
350	1.100	233.5	21.23	29.6	29.0	0.733
400	0.9620	258.2	26.84	33.0	36.4	0.737
450	0.8554	281.4	32.90	36.3	44.4	0.741
500	0.7698	303.3	39.40	41.2	55.1	0.716
550	0.6998	324.0	46.30	44.1	63.8	0.726
600	0.6414	343.7	53.59	47.3	73.5	0.729
700	0.5498	380.8	69.26	52.8	93.1	0.744
800	0.4810	415.2	86.32	58.9	116	0.743
900	0.4275	447.2	104.6	64.9	141	0.740
1000	0.3848	477.0	124.0	71.0	169	0.733
1100	0.3498	505.5	144.5	75.8	196	0.736
1200	0.3206	532.5	166.1	81.9	229	0.725
1300	0.2960	588.4	188.6	87.1	262	0.721
Water Vapor (Steam)						
380	0.5863	127.1	21.68	24.6	20.4	1.06
400	0.5542	134.4	24.25	26.1	23.4	1.04
450	0.4902	152.5	31.11	29.9	30.8	1.01
500	0.4405	170.4	38.68	33.9	38.8	0.998
550	0.4005	188.4	47.04	37.9	47.4	0.993

600	0.3652	206.7	56.60	42.2	57.0	0.993
650	0.3380	224.7	66.48	46.4	66.8	0.996
700	0.3140	242.6	77.26	50.5	77.1	1.00
750	0.2931	260.4	88.84	54.9	88.4	1.00
800	0.2739	278.6	101.7	59.2	100	1.01
850	0.2579	296.9	115.1	63.7	113	1.02

*Sources:

[1] F.P. Incropera, D.P. DeWitt, T.L. Bergman and A.S. Lavine, "Fundamentals of Heat and Mass Transfer", Sixth Edition, John Wiley & Sons, 2006.

[2] NIST Chemistry WebBook.

[3] Y. Wang and K. S. Chen, "PEM fuel cells: thermal and water management fundamentals". Momentum Press, 2013.

[4] Engineering ToolBox.

Practical Handbook of Thermal Fluid Science, 2023, 211-214

Appendix V

Appendix V. Thermal Properties of Selected Materials*

Thermal Properties of Selected Metallic Solids

Composition	Melting Point (K)	Properties at 300 K / 353K†				Properties at Various Temperatures (K) k (W/m·K) / c_p (J/kg·K)						
		ρ (kg/m³)	c_p (J/kg·K)	k (W/m·K)	$\alpha \cdot 10^6$ (m²/s)	100	200	400	500	600	700	800
Aluminum												
Pure	933	2702	906	237	97.1	302	237	240	237	232	226	220
			901†	240†		485	802	935	996	1042	1091	1149
Alloy 2024-T6 (4.5% Cu, 1.5% Mg, 0.6% Mn)	775	2770	875	177	73.0	65	163	186		186		
						473	787	925		1042		
Alloy 195, Case (4.5% Cu)		2790	883	168	68.2		174			185		
							-			-		
Copper												
Pure	1358	8933	386	398	117	483	413	392	388	383	377	371
			398†	394†		252	356	400	404	414	423	438
Commercial bronze (90% Cu, 10% Al)	1293	8800	420	52	14		42	52		59		
							785	460		545		
Phosphor gear bronze (89% Cu, 11% Sn)	1104	8780	355	54	17		41	65		74		
							-	-		-		
Catridge brass (70% Cu, 30% Zn)	1188	8530	380	110	33.9	75	95	137		149		
							360	395		425		
Constantan (55% Cu, 45% Ni)	1493	8920	384	23	6.71	17	19					
						237	362					
Iron												
Pure	1810	7870	443	80.3	23.1	132	94.0	69.4	61.3	54.7	48.7	43.3
			441†	74.1†		216	385	486	495	566	619	686
Armco (99.75% pure)		7870	447	72.7	20.7	95.6	80.6	65.7		53.1		42.2
						215	384	490		574		680
Carbon steels												
Plain carbon (Mn ≤ 1%, Si ≤ 0.1%)		7854	434	60.5	17.7			56.7		48.0		39.2
								487		559		685
AISI 1010		7832	434	63.9	18.8			58.7		48.8		39.2
								487		559		685
Carbon-silicon (Mn ≤ 1%, 0.1% < Si ≤ 0.6%)		7817	446	51.9	14.9			49.8		44.0		37.4
								501		582		699
Carbon-manganese-silicon (1% < Mn ≤ 1.65%, 0.1% < Si ≤ 0.6%)		8131	434	41.0	11.6			42.2		39.7		35.0
								487		559		685

Stainless steels

AISI 302		8055	480	15.1	3.91			17.3		20.0		22.8
								512		559		585
AISI 304	1670	7900	477	14.9	3.95	9.2	12.6	16.6		19.8		22.6
						272	402	515		557		582
AISI 316		8238	468	13.4	3.48			15.2		18.3		21.3
								504		550		576
AISI 347		7978	480	14.2	3.71			15.8		18.9		21.9
								513		559		585
Nickel												
Pure	1728	8900	444	90.5	23.0	158	106	80.1	72.1	65.5	65.3	67.4
			461†	84.7†		232	383	477	527	590	524	524
Nichrome	1672	8400	420	12	3.4			14		16		21
(80% Ni, 20% Cr)								480		525		545
Inconel X-750	1665	8510	439	11.7	3.1	8.7	10.3	13.5		17.0		20.5
(73% Ni, 15% Cr,						-	372	473		510		546
6.7% Fe)												
Platinum												
Pure	2045	21450	132	71.4	25.1	77.5	72.4	71.6	72.2	73.0	74.1	75.5
			134†	71.5†		100	125	136	139	141	144	146
Alloy 60Pt-40Rh	1800	16630	162	47	17.4			52		59		65
(60% Pt, 40% Rh)								-		-		-
Silicon	1685	2330	712	148	89.2	884	264	98.9	76.2	61.9	50.8	42.2
			754†	118†		259	557	785	831	852	869	886
Silver	1235	10500	237	427	174	450	430	420	413	405	397	389
			237†	423.8†		187	225	239	243	248	253	258
Tungsten	3660	19300	133	178	68.3	235	197	162	149	139	133	128
			136†	169.5†		87	122	136	138	140	143	145

†: **Properties at 353K**

Thermal Properties of Selected Nonmetallic Solids

| Composition | Melting Point (K) | Properties at 300 K / 353K | | | | Properties at Various Temperatures (K) | | | | |
| | | ρ (kg/m³) | c_p (J/kg·K) | k (W/m·K) | $\alpha \cdot 10^6$ (m²/s) | \multicolumn{5}{c}{k (W/m·K) / c_p (J/kg·K)} |
						100	200	400	600	800
Carbon Amorphous	1500	1950	-	1.60	-	0.67	1.18	1.89	2.19	2.37
						-	-	-	-	-
Diamond, type IIa insulator	-	3500	516	2000 1831[†]	-	9,800 36	4,300 197	1540 853	1344	1626
Graphite, pyrolytic	2273	2210								
k, ‖ to layers				2000 1606[†]		4980	3250	1460	930	680
k, ⊥ to layers				9.5		39	15	7.0	4.4	3.2
c_p			709	4.73[†]		136	411	992	1406	1650
Graphite fiber epoxy (25% vol) composite	450	1400								
k, heat flow ‖ to fibers				11.1		5.7	8.7	13.0		
k, heat flow ⊥ to fibers				0.87		0.46	0.68	1.1		
c_p			935			337	642	1216		
Nafion[‡] Water level 3 Water level 22		2000		0.185 0.25						
Sulfur	392	2070	704 753[†]	0.206 0.216[†]	0.141	0.165 403	0.185 606			

[†]: **Properties at 353K**

[‡]: **Water lever = H_2O/HSO_3**

Thermal Properties of Common Solids

Description/ Composition	Temperature (K)	Density, ρ (kg/m^3)	Thermal Conductivity, k (W/m·K)	Specific Heat, c_p (J/kg·K)
Ice	273	920	1.88	2040
	263		2.48	
	253	-	2.57	1945
	243		2.68	
Paper	300	930	0.180	1340
	353	-	0.186	-
Teflon	300	2200	0.35	1050
	353		0.41	
	400		0.45	-
Polycarbonate	296	1200	0.29	1250
Soil	300	2050	0.52	1840
Sand	300	1515	0.27	800
	353		0.27	
Snow	273	110	0.049	-
		500	0.190	-
Glass				
Pyrex (borosilicate)	60-100	2210	1.3	753
Rock				
Granite, Barre	300	2630	2.79	775
Limestone, Salem	300	2320	2.15	810
Limestone, Indiana	373	2300	1.1	900
Marble, Halston	300	2680	2.80	830
Quartzite, Sioux	300	2640	5.38	1105
Sandstone, Berea	300	2150	2.90	745

Sources:

[1] F.P. Incropera, D.P. DeWitt, T.L. Bergman and A.S. Lavine, "Fundamentals of Heat and Mass Transfer", Sixth Edition, John Wiley & Sons, 2006.

[2] Y. S. Touloukian and C. Y. Ho, Eds. "Thermophysical properties of Matter", Vol.1; Vol. 2; Vol. 4; Vol. 5, CRC Handbook of Chemistry and Physics, Cleveland, Ohio : CRC Press, 1977.

[3] J.H. Lienhard IV and J.H. Lienhard V, "A Heat Transfer Textbook", 3rd Ed., Phlogiston Press.2008

[4] V.F. Petrenko and R.W. Whitworth, "Physics of Ice", Oxford University Press, 2003.

[5] Y. Wang and K. S. Chen, "PEM fuel cells: thermal and water management fundamentals". Momentum Press, 2013.

This appendix presents the analytical solutions to a series of problems of heat conduction with thermal energy generation for one-dimensional (1-D) geometry under steady-state condition. The form of the heat equation differs, according to the object shape: a plane wall, a cylindrical shell, and a spherical shell. In each shape, several options are available to set the boundary conditions, determined by the real problem. Each set of boundary conditions results in specific solutions of the temperature distribution and heat rate (or heat flux).

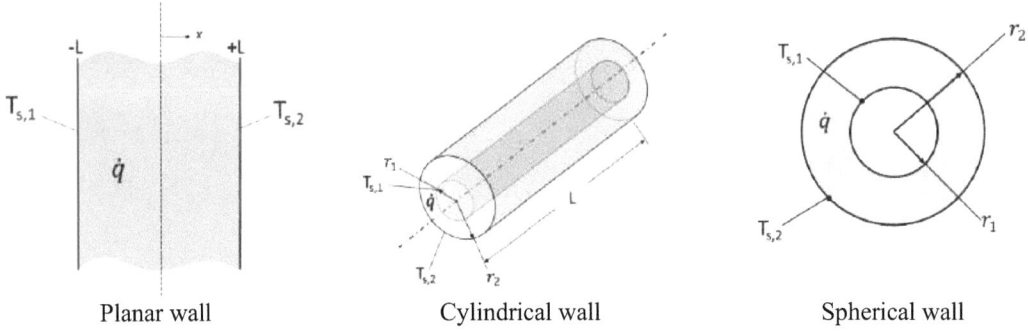

| Planar wall | Cylindrical wall | Spherical wall |

Figure A6.1. 1-D heat conduction with uniform heat generation: a plane wall with asymmetric surface conditions, a cylindrical shell, and a spherical shell.

For the three geometries in Figure **A.61** with uniform temperature $T_{s,1}$ and $T_{s,2}$ at the two surfaces, solutions are readily obtained :

1-D, Steady-State Solutions to the Heat Equations for Plane, Cylindrical, and Spherical Walls with Uniform Heat Generation and Asymmetrical Surface Conditions

Temperature Distribution

Plane Wall
$$T(x)=\frac{\dot{q}L^2}{2k}\left(1-\frac{x^2}{L^2}\right)+\frac{T_{s,2}-T_{s,1}}{2}\frac{x}{L}+\frac{T_{s,1}+T_{s,2}}{2}$$

Cylindrical Wall
$$T(r)=T_{s,2}+\frac{\dot{q}r_2^2}{4k}\left(1-\frac{r^2}{r_2^2}\right)-\left[\frac{\dot{q}r_2^2}{4k}\left(1-\frac{r_1^2}{r_2^2}\right)+\left(T_{s,2}-T_{s,1}\right)\right]\frac{\ln(r_2/r)}{\ln(r_2/r_1)}$$

Spherical Wall
$$T(r)=T_{s,2}+\frac{\dot{q}r_2^2}{6k}\left(1-\frac{r^2}{r_2^2}\right)-\left[\frac{\dot{q}r_2^2}{6k}\left(1-\frac{r_1^2}{r_2^2}\right)+\left(T_{s,2}-T_{s,1}\right)\right]\frac{(1/r)-(1/r_2)}{(1/r_1)-(1/r_2)}$$

Heat Flux

Plane Wall
$$q''(x)=\dot{q}x-\frac{k}{2L}\left(T_{s,2}-T_{s,1}\right)$$

Cylindrical Wall	$q''(r) = \dfrac{\dot{q}r}{2} - \dfrac{k\left[\dfrac{\dot{q}r_2^2}{4k}\left(1-\dfrac{r_1^2}{r_2^2}\right) + \left(T_{s,2}-T_{s,1}\right)\right]}{r\ln\left(r_2/r_1\right)}$
Spherical Wall	$q''(r) = \dfrac{\dot{q}r}{3} - \dfrac{k\left[\dfrac{\dot{q}r_2^2}{6k}\left(1-\dfrac{r_1^2}{r_2^2}\right) + \left(T_{s,2}-T_{s,1}\right)\right]}{r^2\left[\left(1/r_1\right)-\left(1/r_2\right)\right]}$

Heat Rate

Plane Wall	$q(x) = \left[\dot{q}x - \dfrac{k}{2L}\left(T_{s,2}-T_{s,1}\right)\right]A_x$
Cylindrical Wall	$q(r) = \dot{q}\pi Lr^2 - \dfrac{2\pi Lk}{\ln\left(r_2/r_1\right)}\cdot\left[\dfrac{\dot{q}r_2^2}{4k}\left(1-\dfrac{r_1^2}{r_2^2}\right) + \left(T_{s,2}-T_{s,1}\right)\right]$
Spherical Wall	$q(r) = \dfrac{\dot{q}4\pi r^3}{3} - \dfrac{4\pi k\left[\dfrac{\dot{q}r_2^2}{6k}\left(1-\dfrac{r_1^2}{r_2^2}\right) + \left(T_{s,2}-T_{s,1}\right)\right]}{\left(1/r_1\right)-\left(1/r_2\right)}$

In practice, a uniform surface heat flux or a convection boundary condition is also frequently encountered. The third kind is applied by using the overall heat transfer coefficient U in lieu of the convection coefficient h.

Surface Heat Flux Conditions and Energy Balance for 1-D, Steady-State Solutions to the Heat Equation for Plane, Cylindrical, and Spherical Walls with Uniform Generation

Plane Wall

Uniform Surface Heat Flux

$$x = -L: \quad q''_{s,1} = -\dot{q}L - \frac{k}{2L}\left(T_{s,2}-T_{s,1}\right)$$

$$x = +L: \quad q''_{s,2} = \dot{q}L - \frac{k}{2L}\left(T_{s,2}-T_{s,1}\right)$$

Prescribed Transport Coefficient and Fluid Flow Temperature

$$x = -L: \quad U_1\left(T_{\infty,1}-T_{s,1}\right) = -\dot{q}L - \frac{k}{2L}\left(T_{s,2}-T_{s,1}\right)$$

$$x = +L: \quad U_2\left(T_{s,2}-T_{\infty,2}\right) = \dot{q}L - \frac{k}{2L}\left(T_{s,2}-T_{s,1}\right)$$

Cylindrical Wall

Uniform Surface Heat Flux

$r = r_1:$

$$q_{s,1}'' = \frac{\dot{q}r_1}{2} - \frac{k\left[\frac{\dot{q}r_2^2}{4k}\left(1 - \frac{r_1^2}{r_2^2}\right) + (T_{s,2} - T_{s,1})\right]}{r_1 \ln(r_2/r_1)}$$

$r = r_2:$

$$q_{s,2}'' = \frac{\dot{q}r_2}{2} - \frac{k\left[\frac{\dot{q}r_2^2}{4k}\left(1 - \frac{r_1^2}{r_2^2}\right) + (T_{s,2} - T_{s,1})\right]}{r_2 \ln(r_2/r_1)}$$

Prescribed Transport Coefficient and Fluid Flow Temperature

$r = r_1:$

$$U_1(T_{\infty,1} - T_{s,1}) = \frac{\dot{q}r_1}{2} - \frac{k\left[\frac{\dot{q}r_2^2}{4k}\left(1 - \frac{r_1^2}{r_2^2}\right) + (T_{s,2} - T_{s,1})\right]}{r_1 \ln(r_2/r_1)}$$

$r = r_2:$

$$U_2(T_{s,2} - T_{\infty,2}) = \frac{\dot{q}r_2}{2} - \frac{k\left[\frac{\dot{q}r_2^2}{4k}\left(1 - \frac{r_1^2}{r_2^2}\right) + (T_{s,2} - T_{s,1})\right]}{r_2 \ln(r_2/r_1)}$$

Spherical Wall

Uniform Surface Heat Flux

$r = r_1:$

$$q_{s,1}'' = \frac{\dot{q}r_1}{3} - \frac{k\left[\frac{\dot{q}r_2^2}{6k}\left(1 - \frac{r_1^2}{r_2^2}\right) + (T_{s,2} - T_{s,1})\right]}{r_1^2\left[(1/r_1) - (1/r_2)\right]}$$

$r = r_2:$

$$q_{s,2}'' = \frac{\dot{q}r_2}{3} - \frac{k\left[\frac{\dot{q}r_2^2}{6k}\left(1 - \frac{r_1^2}{r_2^2}\right) + (T_{s,2} - T_{s,1})\right]}{r_2^2\left[(1/r_1) - (1/r_2)\right]}$$

Prescribed Transport Coefficient and Fluid Flow Temperature

$r = r_1:$

$$U_1(T_{\infty,1} - T_{s,1}) = \frac{\dot{q}r_1}{3} - \frac{k\left[\frac{\dot{q}r_2^2}{6k}\left(1 - \frac{r_1^2}{r_2^2}\right) + (T_{s,2} - T_{s,1})\right]}{r_1^2\left[(1/r_1) - (1/r_2)\right]}$$

$r = r_2:$

$$U_2(T_{s,2} - T_{\infty,2}) = \frac{\dot{q}r_2}{3} - \frac{k\left[\frac{\dot{q}r_2^2}{6k}\left(1 - \frac{r_1^2}{r_2^2}\right) + (T_{s,2} - T_{s,1})\right]}{r_2^2\left[(1/r_1) - (1/r_2)\right]}$$

In the foregoing configurations, a plane wall with one adiabatic surface, a solid cylinder (a circular rod), and a sphere, as shown in Figure **A6.2**, may be encountered. The corresponding boundary condition at the adiabatic surface follows $dT/dx|_{x=0}=0$ or $dT/dr|_{r=0}=0$. The solutions are based on setting a uniform temperature T_s at $x=L$ and $r=r_0$. If T_s is not known, it may be determined by the energy balance equation at the surface.

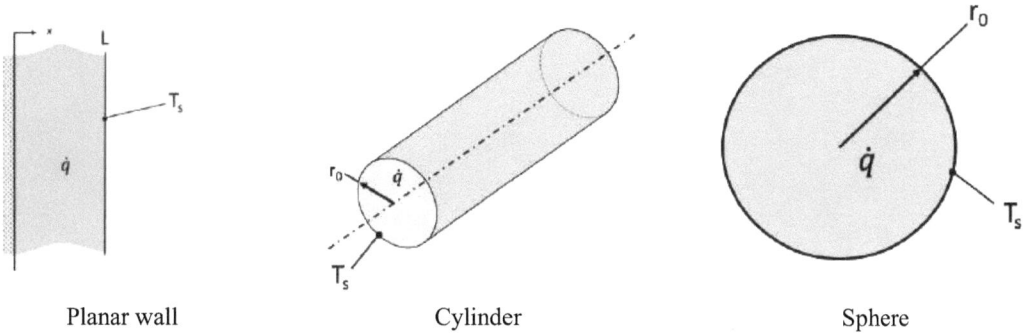

| | Planar wall | Cylinder | Sphere |

Figure A6.2. 1-D conduction systems with uniform thermal energy generation: a plane wall with one adiabatic surface, a cylindrical rod, and a sphere.

1-D, Steady-State Solutions to the Heat Equation for Uniform Generation in a Plane Wall, a Solid Cylinder, and a Solid Sphere with One Adiabatic Surface.

Temperature Distribution

Plane Wall
$$T(x)=\frac{\dot{q}L^2}{2k}\left(1-\frac{x^2}{L^2}\right)+T_s$$

Circular Rod
$$T(r)=\frac{\dot{q}r_o^2}{4k}\left(1-\frac{r^2}{r_o^2}\right)+T_s$$

Sphere
$$T(r)=\frac{\dot{q}r_o^2}{6k}\left(1-\frac{r^2}{r_o^2}\right)+T_s$$

Heat Flux

Plane Wall $q''(x)=\dot{q}x$

Circular Rod $q''(r)=\frac{\dot{q}r}{2}$

Sphere $q''(r)=\frac{\dot{q}r}{3}$

Heat Rate

Plane Wall $q(x)=\dot{q}xA_x$

Circular Rod $q(r)=\dot{q}\pi Lr^2$

Sphere $q(r)=\frac{\dot{q}4\pi r^3}{3}$

*Sources:

[1] F.P. Incropera, D.P. DeWitt, T.L. Bergman and A.S. Lavine, "Fundamentals of Heat and Mass Transfer", Sixth Edition, John Wiley & Sons, 2006

[2] Y. Wang and K. S. Chen, "PEM fuel cells: thermal and water management fundamentals". Momentum Press, 2013.

UCIrvine
University of California, Irvine

Standard Operating Procedure

ENTER TITLE HERE

Department:	
Completion Date:	
Approval (by PI / Lab Manager) Date:	
Principal Investigator:	
Principal Investigator Signature:	
Internal Lab Safety Coordinator/Lab Manager:	
Lab Phone:	
Office Phone:	
Emergency Contact:	*(Name and Phone Number)*
Location(s) covered by this SOP:	*(Building/Room Number)*

Type of SOP: ☐ Process ☐ Hazardous Chemical ☐ Experiment ☐ Equipment Use

Contents

Purpose and Scope
Responsibility
Materials and Equipment
Definitions
Specific Safety and Environmental Hazards
Hazard Control
Location of nearest emergency safety equipment
Shipping and Receiving Requirements
Step-by-step Operating Procedure
Special handling procedures, transport, and storage requirements
Preventive Maintenance

Monitoring and Safety Systems
Waste Disposal/Cleanup
Emergency Response Plan Procedure
References
Preventive maintenance
Monitoring and Safety Systems
Emergency Response Plan
References
Training Requirement
Additional Notes and Attachments
Documentation of Training

Title of SOP

UCIrvine
University of California, Irvine

Read and review any applicable manufacturer/vendor safety information before developing standard operating procedure and performing work

*** NOTE: Each section needs to be complete with clear and detailed information based on the blue/italic font instruction. SOP must be approved and dated by the PI or lab supervisor.

1. **Purpose and Scope of Work/Activity:**

This section identifies the goal of the SOP to answer why the SOP is being written. It needs to be detailed enough so that the intended user can recognize what the document covers. Briefly summarize the process including an estimate of how long the process takes and how frequent it will be conducted.

The scope section identifies who needs to follow the procedure and what the procedure covers. This allows everyone to have the same starting point. You can also add a photo of your equipment.

2. **Responsibility**

Identify the personnel that have a primary roles in the SOP and describe how their responsibilities relates to this SOP. If necessary, include contact information.

3. **Materials and Equipment**

List all of the materials needed to complete the procedure. If the procedure is written to operate a specific piece of equipment, make sure the user guide for the machine is listed in the Reference section, and that users have been trained on operating the equipment prior to performing the procedure.

List all of the materials needed to complete the procedure. List conditions that pose a hazard such as extreme temperature, elevated pressure, reduced pressure, etc. and list the potential hazards

4. **Definitions**

In this section, define any acronyms or abbreviations that are used in the procedure.

5. **Specific Safety and Environmental Hazards:**

List any warning or precautions before performing a procedure.

6. **Hazard Control(s)**
   7.1. **Engineering/Ventilation Controls**

SOP template   Revised: date

Title of SOP

UCIrvine
University of California, Irvine

In this section describe any specific engineering controls which are required to prevent operator exposures to hazards such as fume hood, interlocks on equipment, explosion shielding, ultraviolet light shielding, and safety features on equipment.

7.2. Administrative Controls

In this section describe any specific administrative controls which are required to prevent operator exposures to hazards such as

> *Proper labeling and storage*
> *Good laboratory housekeeping*
> *Testing equipment function and inspecting for damage prior to use*
> *Shift-hand-off procedure,*
> *Equipment status reporting*
> *Mothballing process, etc.*

7.3. Personal Protective Equipment

State the personal protective equipment selected and when it's required, and specific hygiene practices if needed. For example: Safety glasses, nitrile gloves, cryo gloves, absorbent bench paper, respiratory protection, lab coat and body protection, etc.

7. Location of nearest emergency safety equipment

ITEMS	

8. Shipping and Receiving Requirements

Describe shipping or receiving requirement, especially for highly toxic, highly reactive, unstable, highly flammable and corrosive materials.

9. Step-by-step Operating Procedure

Provide the steps required to perform this procedure.

Title of SOP

For a process*: Write enough detailed description of the procedure to guide the user through the process including details of startup, normal condition operation, temporary operation condition and emergency shut-down, etc.*

Also cover enough information as following:

1. *Chemical concentrations, gas amount*
2. *Pressure limits, temperature ranges*
3. *Flow rates*
4. *Special safety equipment is to be utilized.*
5. *Schematics or pictures for complex setups.*
6. *Highlight safety precautions put in place*
7. *What to do when an upset condition occurs*
8. *What alarms and instruments are pertinent if an upset condition occurs*
9. *If lockout/tagout is required*

Some tips:

- *Use numbers for steps and sub-steps that have to be performed in a specific sequence*
- *Use bullets for steps or items that can be performed in any order.*
- *Solicit ideas for other users.*
- *Include a flow diagram to help interpret more complex procedures.*
- *Include pictures and label different components.*

For Equipment: *Describe the step-by-step procedure for using the equipment properly. Include details for potential equipment failure if something is done improperly in the procedure. Describe how to power down the equipment at the end of use.*

10. Special handling procedures, transport, and storage requirements

Describe special handling and storage requirements for hazardous chemicals/gases in the laboratory, especially for highly reactive, unstable and highly flammable materials and corrosives. Describe transport and secondary containment requirement, between the laboratories or between facilities.

11. Preventive Maintenance

Clean up and preventive maintenance is important for keeping equipment in safe working order. In this section, any regular maintenance and / or calibration frequency for research equipment, instrumentation and/or facilities should be included here.

12. Monitoring and Safety Systems

This section includes a list of all monitoring systems such as gas detectors, safety interlocks, equipment guards, fail safe control logic, etc. noted.

13. Waste Disposal/Cleanup

Describe waste handling procedures for collecting, storing, and disposing of liquid, solid, or mixed waste (including associated contaminated debris) generated in this procedure. Include container types, labeling, segregation, and any required treatments prior to disposal (e.g. neutralization or decontamination steps).

SOP template   Revised: date

Title of SOP

UCIrvine
University of California, Irvine

For more assistant please contact EH&S.

14. Emergency Response Plan

In this section describe any special procedure for spills, releases or fire. Indicate how accidental events should be handled and by whom. List emergency contact number.

First Aid Procedure

If inhaled

1. Move to fresh air
2. Have victim rest in half-upright position
3. Seek medical attention immediately

In case of Skin Contact

1. Immediately flush skin with plenty of water for at least 15 minutes
2. Remove contaminated clothing and shoes
3. Get medical attention immediately

In case of eye contact

1. Immediately flush eyes with plenty of water for at least 15 minutes from emergency eyewash station
2. Get medical attention immediately

If swallowed

1. Do not induce vomiting unless directed to do so by medical personnel
2. Never give anything by mouth to an unconscious person.
3. Loosen tight clothing such as a collar, tie, belt or waistband.
4. Get medical attention immediately

Life-threatening emergencies (Such as: fire, explosion, large-scale spill or release, compressed gas leak, valve failure, etc)

1. Evacuate the room and close the door behind you
2. Secure the room to prevent entry
3. Alert people in the area and activate the local alarm systems
4. **Call 911 – Tell the dispatcher the name of the gas or chemical.**
5. Provide local notification
6. Report to EH&S at x46200 within 8 hours
7. Complete online incident report at *https://www.ehs.uci.edu/apps/hr/index.jsp*

Identify the area management staff that must be contacted and include their work and home numbers. This must include the PI and may include the safety coordinator and facility manager.

In case personnel exposed or injured

1. Remove the victim from the area if it is safe to do so
2. Follow first air protocol as mentioned above

Title of SOP

UCIrvine
University of California, Irvine

3. Provide safety data sheets (SDSs) for all chemicals to Emergency Medical Technician (EMT) or to the hospital
4. Report to EH&S x46200
5. Complete the online incident form *https://www.ehs.uci.edu/apps/hr/index.jsp* or Human Resources, Workers Compensation at x9152

Non-life threatening emergencies

1. Notify your supervisor or faculty staff
2. Report to

Identify the area management staff that must be contacted and include their work and home numbers. This must include the PI and may include the safety coordinator and facility manager.

For spill & accident procedure

In the event of a small spill or release that can be cleaned by a trained local personnel follow below steps:

1. Use appropriate personal protective equipment and clean up material for chemical spilled
2. Double bag spill waste in clear plastic bags, label and take to the next chemical waste pick-up

In case of large spill or release:

1. Evacuate the spill area
3. Post someone or mark-off hazardous area with tape and warning signs
4. **Call 911** and EH&S at x46200 for assistance
5. Keep the fire extinguisher nearby
   Note: Fire extinguishers containing water are not suitable for flammable liquid fires

Building maintenance emergencies (for example: power outage, plumbing leaks)

Submit a Facilities Service Request (*https://service.fac.uci.edu/html/en/default/reportTemplate/viewPageReport.jsp*) or call appropriate building manager.

Additional emergency procedures: Describe additional, local emergency procedures.

15. References

This section should include the references that were used to produce this SOP.

- Online SDSs can be accessed at http://www.ehs.uci.edu/msds.html

16. Training Requirement

In this section list the general and laboratory specific training required.

17. Additional Notes and Attachments

In this section list any notes or attachments needed to implement this SOP.

SOP template

Revised: date

Title of SOP

UCIrvine
University of California, Irvine

18. Documentation of Training
- Any deviation from this SOP requires approval from PI.
- Prior to conducting any work with the equipment, designated personnel must provide training to his/her laboratory personnel specific to the hazards and procedures involved in working with this process.
- The Principal Investigator must provide his/her laboratory personnel with a copy of this SOP and a copies of any SDS provided by the manufacturer for any chemicals used.
- The Principal Investigator must ensure that his/her laboratory personnel have attended appropriate laboratory safety training or refresher training annually.

Title of SOP

UCIrvine
University of California, Irvine

I have read and understand the content of this SOP:

Name			Date

SUBJECT INDEX

A

Air 64, 67, 102, 109, 118, 155, 156, 160, 168, 172, 174, 175
  compression 64
  compressors 67
  conditioners 102, 160, 172, 174, 175
  cycle machine 168
  enthalpy 109
  injection rate 118, 155, 156
Aircrafts 121, 124, 126, 130, 168
  flying 130
  powered 168
Airplane flying 135
Analysis 105, 146, 179
  engineering 179
  open-system 146
  thermodynamic system 105
Applications 1, 3, 12, 22, 28, 66, 70, 81, 84, 93, 122, 123, 138, 148
  industrial 1
  renewable energy 66
  waste 84
Atmospheric pressure 144

B

Bernoulli constant 126
BFW pressure 118
Bimetallic 3, 38
  devices 3
  thermometers 38
Boiler(s) 60, 61, 62, 69, 70, 72, 73, 74, 81, 102, 103, 104, 106, 107, 108, 109, 117, 118
  commercial 106
  efficiency measures 106
  feed water (BFW) 70, 72, 73, 81, 106, 108
  functions 109
  natural-gas-fueled 104
  waste heat 109
  water-tube 61

Brake 144, 150
  hysteresis 150
  magnetic powder 150
Brayton cycle processes 63
Burning fossil fuels 106

C

Calibration 16
  constant 16
  equation 16
Calorically perfect gas (CPG) 140, 141
Chemical energy 102, 104, 108, 150
Chilled water return (CWR) 70, 76, 77, 78
Chiller and boiler building 73
Combined-cycle configuration 60
Combustion 64, 68, 71, 104, 106, 113, 139, 140, 141, 142, 147, 148, 154, 155, 156
  fuel-rich 155
  energy 148
  expansion strokes 155
  process 142, 147
  turbine generator (CTG) 68, 71
  turbines 64, 113
Commercial devices 39
Compressed 64, 138, 161, 166
  air 64, 138, 166
  natural gas plants 161
Compression 121, 142, 148, 154
  ratio 142, 148, 154
  shocks 121
  stroke 148
Compressor 63, 64, 66, 70, 143, 161, 163, 169, 170, 171, 174
  air-conditioning 143
  drive 70
  irreversible 174
Constant 61, 64, 171
  enthalpy process 171
  pressure heat addition 61
  pressure heat rejection 64

www.ingramcontent.com/pod-product-compliance
Lightning Source LLC
Chambersburg PA
CBHW050828220326
41598CB00006B/333

A Guide to Balance and Dizziness

The second edition of *A Guide to Balance and Dizziness* by Charles M. Plishka returns to address the increasing number of referrals to treat balance impairment, gait disorders, and dizziness. This expanded edition offers tests, measures, and interventions aligned to available research studies to support evidence-based practice. By gaining a fundamental understanding of how we balance, recognizing and understanding the signs and symptoms of patients will become much easier.

Along with numerous updated diagrams and photos, the new edition also comes with access to a website containing video clips that demonstrate key evaluation and treatment techniques. The result will be a better evaluation, treatment plan, and outcome.

Topics and Features Include

- Tests to evaluate the balance-impaired patient
- Tests and interventions for conditions such as benign paroxysmal positional vertigo (BPPV), vestibular loss, and the central and peripheral causes of dizziness
- Therapy treatments
- "How to" instructions throughout
- Instructor resources for each chapter

A Guide to Balance and Dizziness is a user-friendly reference, ideal for professionals involved in assessing and treating balance impairments and dizziness. It is an instructional text for physical therapy students and clinicians and an excellent resource for established physicians, optometrists, vestibular and balance therapy specialists, occupational therapists, nurse practitioners, physician assistants, audiologists, athletic trainers, and other disciplines that evaluate and offer treatment interventions for these conditions.